Mathematics Study Resources

This series comprises direct translations of successful foreign language titles, especially from the German language.

Powered by advances in automated translation, these books draw on global teaching excellence to provide students and lecturers with diverse materials for teaching and study.

Olaf Manz

Encrypt, Sign, Attack

A compact introduction to cryptography

Olaf Manz
Worms, Germany

ISSN 2731-3824 ISSN 2731-3832 (electronic)
Mathematics Study Resources
ISBN 978-3-662-66014-0 ISBN 978-3-662-66015-7 (eBook)
https://doi.org/10.1007/978-3-662-66015-7

This Springer imprint is published by the registered company Springer-Verlag GmbH, DE, part of Springer Nature.
The registered company address is: Heidelberger Platz 3, 14197 Berlin, Germany

Preface

Have you ever wondered whether mobile phones can be used to confide even the most secret secrets? Or whether online banking is really secure these days? Or whether an electronic signature on contracts sent by e-mail meets legal requirements? All of this has something to do with the **encryption** – or **ciphering** – of data, which is sent or stored on data carriers every day in large and ever-increasing quantities via data highways or "wireless".

Textbooks and reference books take a more scientific approach to the topic of data encryption under the title of **cryptography.** They deal with the mathematical theories of the common procedures, describe their algorithms and program-technical realizations, and also deal with many topics of the organizational implementation. As a basis for lectures or seminars, it must in the first instance be the goal to introduce students to scientific work and to introduce them to areas of current research. Practitioners working in the subject also need a correspondingly comprehensive presentation. On the other hand, there are also numerous popular science publications that aim at a generally understandable level. This works very well in this case, since simple ciphering methods can easily be brought to the attention of interested laymen and can be substantiated with examples from everyday practice. The mathematics behind it, however, usually remains hidden.

This book aims to be a balancing act between the two. It is a fact that cryptography can be understood quite comprehensively with very little mathematics. Our goal is therefore, without a theoretical superstructure, to deal specifically with the most important procedures of **encryption, signing** and **authentication,** and to present them in a compact and mathematically understandable manner, which is reflected in many practical examples.

- We focus first on symmetric ciphers, where anyone who knows the cipher procedure can decode it. The procedures go back to antiquity to the **Caesar cipher,** in which each letter in the alphabet is replaced by the letter three places further down. The **Vigenère cipher** from the sixteenth century does this much more subtly, while more modern methods such as the **Triple DES** (Data Encryption Standard) and especially today's standard method **AES** (Advanced Encryption Standard) are considerably more complex.

- But how is it supposed to work that you can encrypt but can't decrypt even with the help of the biggest and most modern computers? The keyword is **public key.** We will learn about the standard methods: **RSA** relies on the difficulty of decomposing large natural numbers into factors, and **Diffie-Hellman** and **ElGamal** exploit the problem that "discrete logarithms" cannot be computed efficiently enough. Here, we even run into "elliptic curves" with **ECDH.**
- Ciphering would certainly be unnecessary if there were not rogues and especially also professional attackers who would expect political, military or economic advantage from the knowledge of secret data and therefore try to "crack" the encryption. In addition to classical attacks using **statistical analysis** and **Friedman's coincidence index,** we learn about **Pollard's** methods for effectively factorizing large natural numbers to potentially "crack" RSA. Finally, we also attack the "discrete logarithm" with **Baby-Step-Giant-Step** and **Pohlig-Hellman**.
- Particularly fatal, however, is an attack in which an unauthorized person not only passively listens in but also actively engages in the message traffic and changes it in their own way. In this case, the recipient of a message is completely unaware of whether the information received in this form really originates from exactly the sender specified. In order to prevent this situation, **digital signatures** are used, for example the **RSA, DSA** or **ECDSA** procedure, thus giving a **man-in-the-middle attack** no chance.
- Of course, we will always deal with practical applications. Historically interesting are, for example, the **Illuminati** cipher and the **Enigma** machine. The **Internet** with **HTTPS** is perhaps the most prominent modern application for secure data transmission, but wireless **WLAN** networks and the **Bluetooth** radio interface are also well protected today. The **PGP Pretty Good Privacy** method is widely used for **e-mails**, while **mobile communications** with **GSM** are only partially secure against eavesdropping, but those with **UMTS/LTE** are much more secure. Another focus is on **online banking, credit cards** and **Bitcoins.** Finally, **e-passports** with their biometric data are also designed to be forgery-proof. Last but not least, data stored on **hard disks**, and thus **passwords** in particular, must be protected against unauthorized access.

The target audience for this book is basically anyone who is enthusiastic about the topic; in particular, it is also intended as an introduction to more advanced literature. We will have to do relatively little, but nevertheless some mathematics. We will need arithmetic with binary numbers (bits) and with remainders modulo a natural number, as well as an understanding of permutations, both for the conceptual background and for one or the other formal derivations. However, we will build this up piece by piece, with special emphasis on the plausibility of the relationships. So, let's plunge into the adventure – and have fun.

As a guide, here is a brief reading guide for the four chapters of this book in advance:

- Chapter 1 is intended as a "warm up", with an overview of important historical ciphers.
- Chapter 2 examines **symmetric ciphers S** (standard methods: **Triple-DES** and **AES**). These cipher procedures depend on a parameter to be kept secret, the so-called **key k,** with the help of which decryption can also be performed. Thus, participant **T** encrypts a secret message **m** as follows:
  - **S**(**m**, **k**).
- But how does participant **T** deliver the comparatively short key **k** to the authorized recipient of the secret message **m,** also by secret means? Chapter 3 introduces the concept of **public-key ciphers E** (standard methods **RSA, (EC)DH, ElGamal**). In these methods, which however require much more computation time than symmetric ciphers, one cannot decrypt **e** from the knowledge of their **key** alone. Participant **T** therefore encrypts **k** using **E** and sends the concatenation as a whole:
  - **E**(**k**, **e**)|| **S**(**m**, **k**).
- But wait: How can the recipient be sure that the received message really comes from participant **T** in exactly this form? In Chap. 4, this problem is solved by means of **digital signatures sig** (standard methods **RSA, (EC)DSA**). Participant **T** signs the message **m,** more precisely a **digital fingerprint h**(**m**) of **m** (standard procedure **SHA**), and additionally sends the signature **sig**(**h**(**m**)) in the following concatenation:
  - **E**(**k**, **e**)|| **S**(**m** || **sig**(**h**(**m**)), **k**).
- Sometimes, however, a more conventional method, the **checksum MAC**, is used as an alternative to the digital signature (standard procedures **CBC-MAC, HMAC**):
  - **E**(**k**, **e**)|| **S**(**m** || **MAC**(**m**), **k**).
- Before sending a secure message, participant **T** usually has to log on to a system (e.g., mobile phone, computer network, bank server) and legitimize himself. As explained at the end of Chap. 4, this can be done "classically", for example by entering a PIN or password, but also with the aid of modern **public key** procedures and in particular with a **digital signature**:
  - **T** ►... **E**(**k**, **e**)|| **S**(**m** || **sig**(**h**(**m**)), **k**).

Worms, Germany Olaf Manz

Contents

Basics and History

1

1.1 What It Is About: The Scenario

1.1.1 An Initial Overview

In today's information age, unimaginable amounts of information are exchanged between senders and receivers in business, in science, in the public and in private. This can take place via online channels as well as via storage media. There are basically three requirements to be considered as shown in Fig. 1.1. The information should

- be transmitted or stored in the most space-saving and compressed way possible,
- be protected against unwanted interception or even unauthorised modification, and
- arrive without significant loss of information despite random disturbances in the transmission channel or damage to the storage media used.

An online channel can be thought of as computer networks (LAN, Internet, etc.), mobile communications networks or digital television via cable or satellite. Storage media include, for example, hard disk or USB stick.

1.1.2 Sending and Storing Information

Let us first use Fig. 1.2 to clarify the terms and the individual steps in a little more detail.

First, the desired information must be structured and "put on paper". This can be done in German, English or another language and should also be illustrated with some graphics and photos. To do this, you may already be using WORD for text passages and TIFF for graphic formats and have thus already **digitized** your documents.

O. Manz, *Encrypt, Sign, Attack*, Mathematics Study Resources,
https://doi.org/10.1007/978-3-662-66015-7_1

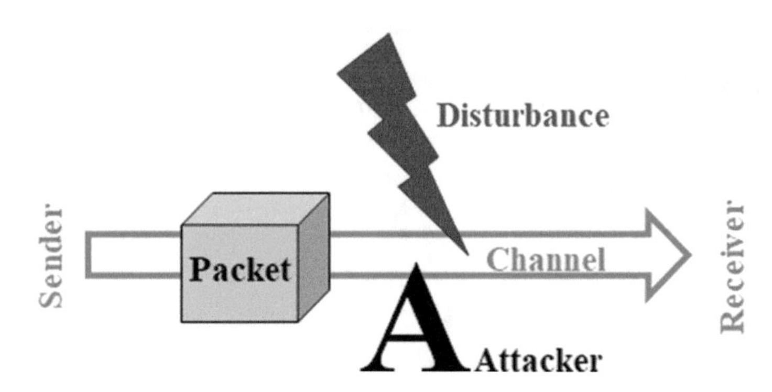

Fig. 1.1 The scenario of data transmission and storage

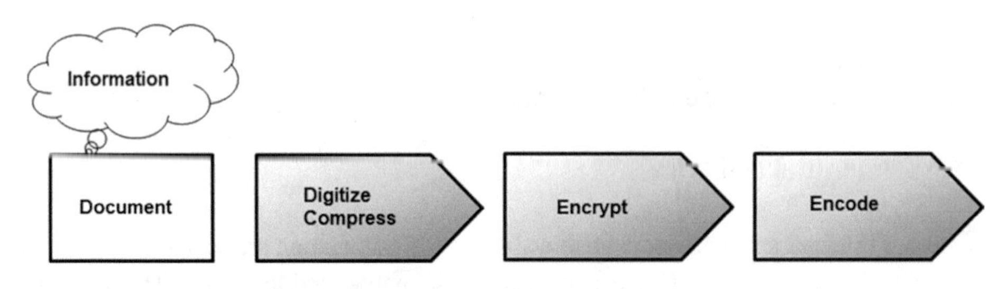

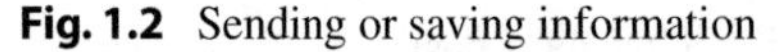

Fig. 1.2 Sending or saving information

When sending or archiving documents, however, it must also be taken into account that the transmission time should be as short as possible and the required storage space as small as possible, i.e. the data should be used in a suitably compressed form. The optimal **digitization** of data together with suitable **compression** is a sub-aspect of the field of **information theory**.

In a further step, our digitized document should be protected against eavesdropping or even changes by unauthorized third parties. To do this, we encrypt its contents in such a way that it cannot be read or even changed by strangers. This is called **ciphering**, the related field is called **cryptography**. In addition, we have to find methods to "crack" the ciphers used, i.e. to put ourselves in the role of an attacker, either in real or virtual terms. This is called **cryptanalysis**.

Last but not least, our transmission channel is susceptible to interference (e.g. short-term noise), or the storage medium used might have been damaged (e.g. scratches on the DVD). In the so-called **encoding** process, we add a little redundant information to our text so that any errors that occur can usually be detected and possibly corrected without asking. This step is the main ingredient of the field of **coding theory** [Man].

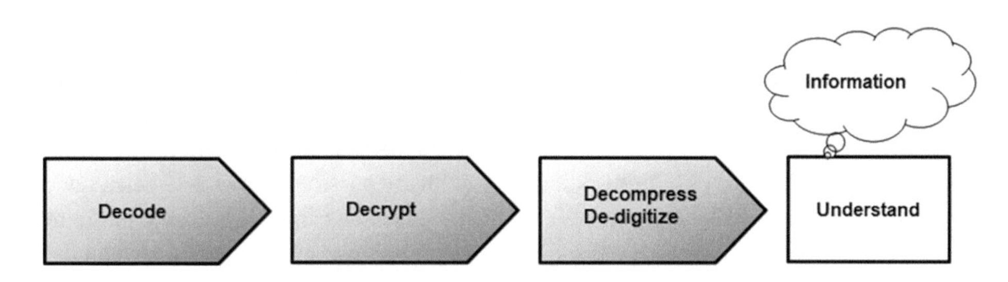

Fig. 1.3 Receiving or reading out information

1.1.3 Receiving and Reading Information

After receiving our document or reading out the corresponding memory contents, the above steps must all be undone, as shown in Fig. 1.3.

Reception errors should at least be detected or, even better, automatically corrected. In addition, the redundancy must be removed again and thus the original message recovered **(decoding)**. Then the message must be decrypted **(deciphering)**, which of course requires that the recipient knows the decryption procedure. Finally, the document must be converted from its compressed and digitized state back into the readable source text including embedded graphics. Only now can the content of the document be understood by the recipient.

The topic of this paper is cryptography and cryptanalysis. Thus, we will discuss ciphering in detail, i.e., how to protect sensitive information against unwanted eavesdropping or even unauthorized modification.

1.2 Alphabets and Digitisation

1.2.1 Alphabets and Modulo Calculation

The basic prerequisite for the transmission or storage of abstract information is first of all its structured documentation. As a rule, a text is documented with letters, and sequences of digits or combinations of digits and letters are used for identifying marks (e.g. passport number), although this does not necessarily have to be digitized at first. However, if you use WORD, for example, the text is automatically digitized. In the case of graphics, which are nowadays no longer created by hand, but e.g. with PowerPoint, as well as digital photos, there is ultimately no other choice anyway. But anyway: The structuring of information is always based on the use of so-called **alphabets**. Here are some examples.

- Alphabet of all capital letters A, B,…, Z
- Alphabet of all digits 0, 1,…, 9
- Alphabet of all digits and capital letters 0, 1,…, 9, A, B,…, Z

For arithmetic implementations, letters naturally have the disadvantage that one cannot calculate with them. But also with the digits a problem arises, because one gets out of their one-digit range very fast when calculating, because for example $7+8=15$ and $3 \bullet 4=12$ are already no simple digits any more. It is better to operate with remainders **modulo** a natural number m instead of letters or digits, i. e. with the possible remainders when dividing by m. Thus the alphabet consists of 0, 1,..., m – 1, which can be added and multiplied modulo m. For example, for m = 10, 7 + 8 has remainder 5 when divided by 10, and thus 7 + 8 = 5, read modulo 10. The product $3 \bullet 4$ has remainder 2 when divided by 10, so $3 \bullet 4 = 2$, also read modulo 10. So sum and product are again an element of the alphabet in our modulo calculation. We can now also calculate with letters by simply taking the m = 26 capital letters as 0, 1, 2,..., 25 and thus as remainders modulo 26.

Calculating modulo a natural number m will turn out to be an important procedure in many places. For two integers, i.e. possibly also negative natural numbers a and b, one writes **a = b (mod m)** and means by this that a and b modulo m are equal, i.e. have the same remainder 0 or 1, or... or m – 1 when divided by m. The following criteria are then equivalent:

- a = b (mod m)
- a – b is divisible by m
- a and b differ only by a multiple of m

Here's how to make that clear:

If a = b (mod m), then a and b have the same remainder r when divided by m. So a and b can be written as $a = q_a \bullet m + r$ and $b = q_b \bullet m + r$. If we subtract the second equation from the first, it follows that $a - b = (q_a - q_b) \bullet m$, and a – b is divisible by m.

If a – b is divisible by m, then a – b can be written as $a - b = q \bullet m$. Therefore $a = b + q \bullet m$ follows, i.e. a and b differ only by a multiple of m.

If a and b differ only by a multiple of m, then a can be written as $a = b + s \bullet m$. We now divide a and b with remainder by m, that is, $a = q_a \bullet m + r_a$ and $b = q_b \bullet m + r_b$ with remainders r_a and r_b in the range 0 to m – 1. Let r_a be greater than or equal to r_b. If we now subtract the second equation from the first and consider $a = b + s \bullet m$, we get $s \bullet m = a - b = (q_a - q_b) \bullet m + (r_a - r_b)$ and consequently $r_a - r_b = (s - q_a + q_b) \bullet m$. Therefore, $r_a - r_b$ is a multiple of m. But since $r_a - r_b$ is also in the range 0 to m – 1, it follows that $r_a - r_b = 0$. Thus, the two residues r_a and r_b are equal, i.e., a = b (mod m).

By far the most important and at the same time simplest alphabet is that of the remainders modulo m = 2, i.e. 0 and 1. This is the so-called **binary** alphabet of **bits**. The addition and multiplication of bits is shown in the following addition and multiplication tables:

+	0	1
0	0	1
1	1	0

.	0	1
0	0	0
1	0	1

It also makes sense to consider alphabets made up of blocks of bits, such as blocks of 2 bits, namely 00, 01, 10, 11, but often blocks of 8 bits are used, as we will now see.

1.2.2 Digitisation and Bytes

So far, we have used the terms digitization and alphabet rather intuitively. In general, **digitization** is understood as the conversion of abstract information or analog values into a sequence of "discrete" characters. As a rule, this involves only a finite number of characters, and their totality is then called an **alphabet**. Against this background, the conversion of a spoken text into a sequence of letters can already be regarded as "digitization". **Digitization** in the narrower – and today always assumed – sense, however, also means that the elements of the alphabet are represented as a **binary string**, i.e. as a sequence of bits 0 and 1. In the case of texts, for example, their letters consist of blocks of bits; in the case of photos and graphics, the same is true for the color and brightness values of the individual image points, so-called **pixels**, as Fig. 1.4 illustrates.

One block length of bits has proven to be particularly useful in the past, namely the length 8, with which one can thus represent up to $2^8 = 256$ characters such as letters, digits or brightness values. A block of 8 bits is called a **byte**. However, the representation of characters is not always limited to one byte and thus to 256 values, but more than 8 bits or even a few bytes can also be used.

There is a 256-character standard that includes upper and lower case letters, numbers, and most special and control characters, called the **ASCII character set** (American

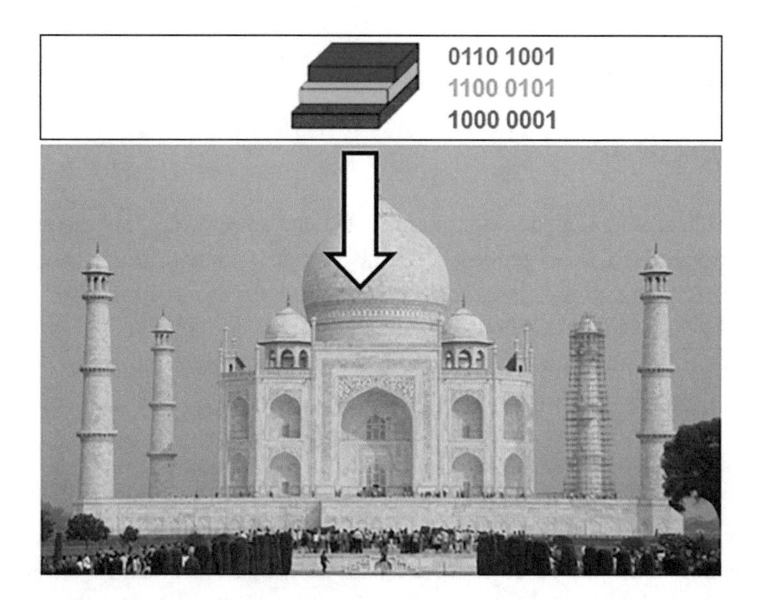

Fig. 1.4 Digitized pixels. (Photo: Olaf Manz)

Table 1.1 Extract from the ASCII table

No.	ASCII	Byte
:::::	:::::	:::::
37	%	0010 0101
38	&	0010 0110
:::::	:::::	:::::
49	1	0011 0001
50	2	0011 0010
:::::	:::::	:::::
65	A	0100 0001
66	B	0100 0010
:::::	:::::	:::::
97	a	0110 0001
98	b	0110 0010
:::::	:::::	:::::

Standard Code for Information Interchange). Although many procedures are more flexible today, people still like to work with digital ASCII characters. In this case, the number of the respective ASCII character must be read as a binary representation in order to derive the desired byte. Table 1.1 shows an extract from the table of all ASCII values as an example.

Historical ciphers, of course, do not yet use bits and bytes, but letters, digits, and possibly some special characters, as we will see in the rest of this chapter.

1.3 Caesar Cipher

1.3.1 Shift Cipher

We begin with historical ciphers in antiquity. Roman sources say that the emperor and general **Gaius Julius Caesar** proceeded for his secret communication in such a way that he replaced each letter of the alphabet by the one three places further, i.e. “A” by “D” and finally “Z” by “C”. Figure 1.5 shows an example.

Therefore, this **cipher** procedure is also called **Caesar cipher**, the underlying alphabet being A, B,…, Z. We first make two observations:

- Of course, instead of three digits, you could have chosen another number i from 0 to 25 and replaced each letter with the one i digits further.
- We have also already learned that instead of letters it is better to use the remainders 0, 1,…, 25 modulo 26. So, based on these observations, the ciphertext for the characters z of our alphabet is $z \rightarrow z + i \pmod{26}$. This is called a **shift cipher**.

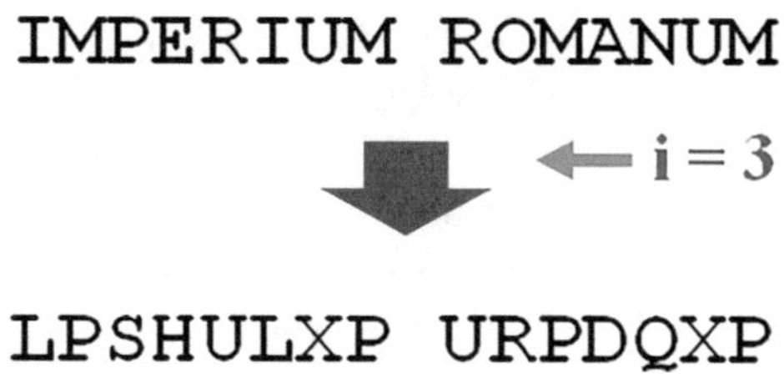

Fig. 1.5 Example of a Caesar cipher

IMPERIUM ROMANUM

$i = 3$

LPSHULXP URPDQXP

But how did the governors in the Roman provinces decode Caesar's orders, i.e. decipher them? Of course, they had to know the shift value $i = 3$, and again they moved all the letters back three places in the alphabet, in which case, for example, "A" becomes "X". Using our algorithmic notation for shift ciphers, this means $z \rightarrow z - i = z + (26 - i) \pmod{26}$. Thus, for the Caesar cipher with $i = 3$ and the letter A, i.e., $z = 0$, we get $0 \rightarrow -3 = 26 - 3 = 23 \pmod{26}$, which is the letter X. In this way, one can unambiguously decipher each character in the encoded letter sequence.

But how secure is a shift cipher? So let's do some cryptanalysis and examine how the encryption method can be cracked. This can be done quite easily by means of statistics. In most languages, the "E" is by far the most frequent letter:

German	17.4%
English	12.7%
French	14.7%
Spanish	13.7%
Italian	12.0%

So you determine in a sufficiently large passage of the ciphertext the most common letter, which could be, for example, the "Q". This then most likely corresponds to the letter "E", which is 12 places before Q in the alphabet. Assuming that the cipher is a shift cipher, one only needs to replace each letter in the ciphertext with the letter 12 places before it in the alphabet, and the readable plaintext is obtained.

1.3.2 Affine Cipher

Now we make our shift cipher a bit more complicated and not only add i, but we also multiply by a j, i.e. $z \rightarrow j \bullet z + i \pmod{26}$. This is called an **affine cipher**. In this case, however, j must be chosen to be coprime to 26, i.e., j odd and j not equal to 13. Only then can $j^{-1} \pmod{26}$ be computed (Sect. 3.1) and the assignment $z \rightarrow j^{-1} \bullet z - j^{-1} \bullet i \pmod{26}$ be used for deciphering. Thus, the receiver must now know both i and j. For example, for $j = 3$, $j^{-1} = 9 \pmod{26}$, since $3 \bullet 9 = 27 = 1 \pmod{26}$.

But even the affine cipher is not much more secure than a shift cipher, because here an unauthorized listener has to count out the second most frequent letter in a ciphertext in addition to the most frequent one. In German, this will correspond to the letters E (i.e.

$z = 4$) and N (i.e. $z = 13$). For example, if one has determined A (i.e., $z = 0$) and F (i.e., $z = 5$) in the ciphertext, then, assuming an affine cipher, $0 = j \bullet 4 + i \pmod{26}$ and $5 = j \bullet 13 + i \pmod{26}$ follows. Subtracting the first equation from the second, we get $5 = j \bullet 9 \pmod{26}$. Now multiplying by 3, we get $15 = j \bullet 9 \bullet 3 = j \bullet 27 = j \pmod{26}$, so $j = 15$. Substituting this into the first equation, we get $0 = 15 \bullet 4 + i = 60 + i = 8 + i \pmod{26}$, so $i = 18$. Therefore, the encryption is $z \rightarrow 15 \bullet z + 18 \pmod{26}$. From this, the deciphering procedure and hence the total plaintext are computable.

1.4 Secret Writing of the Illuminati

1.4.1 Illuminati Alphabet and Secret Writing

A historically popular means of secrecy is that of various secret writings, such as those of the **Illuminati** (Latin for the enlightened), a secret order founded in the eighteenth century, around which numerous myths and conspiracy theories surrounding the Catholic Church are entwined. The Illuminati became famous not least through the best-selling novel of the same name by Dan Brown and the film adaptation starring Tom Hanks. In the Illuminati's cipher [Kuh], the letters and numbers are each replaced by a fixed, self-discovered secret character of the Illuminati alphabet, as Fig. 1.6 shows.

The Illuminati secret writing therefore appears at first glance to be extremely strange and hardly decipherable, as the simple example in Fig. 1.7 shows.

Nevertheless, there are also cryptanalytic starting points here. The frequency of letters is transferred to the uniquely assigned secret character, so that statistical methods can be applied again. The first step is to obtain the statistical frequencies of all letters in as many different languages as possible. Table 1.2 shows some examples.

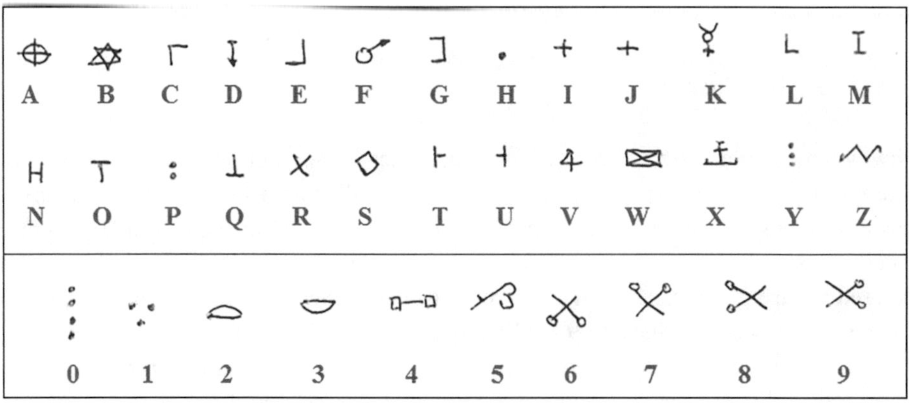

A B C D E F G H I J K L M

N O P Q R S T U V W X Y Z

0 1 2 3 4 5 6 7 8 9

Fig. 1.6 Illuminati alphabet

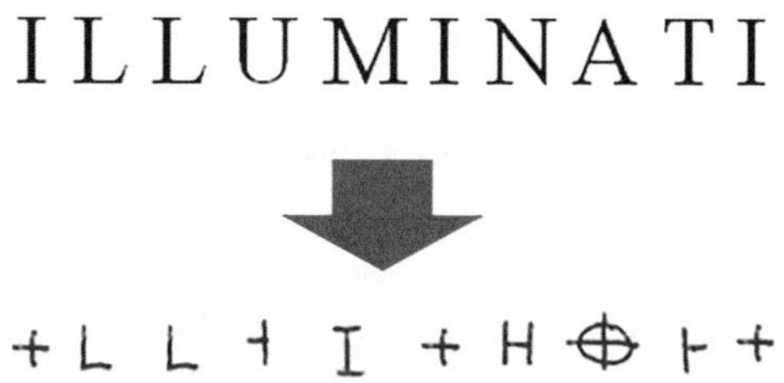

Fig. 1.7 Example of Illuminati secret writing

ILLUMINATI

Table 1.2 Letter frequency in various languages

	German (%)	English (%)	French (%)	Spanish (%)	Italy (%)	Swedish (%)
A	6.50	8.20	7.60	12.50	11.70	9.30
B	1.90	1.50	0.90	1.40	0.90	1.30
C	3.00	2.80	3.30	4.70	4.50	1.30
D	5.10	4.30	3.70	5.90	3.70	4.50
E	17.40	12.70	14.70	13.70	12.00	10.00
F	1.70	2.20	1.10	0.10	0.10	2.00
G	3.00	2.00	0.90	1.00	1.60	3.30
H	4.80	6.10	0.70	0.70	1.50	2.10
I	7.60	7.00	7.50	6.30	11.30	5.10
J	0.30	0.20	0.50	0.40	–	0.70
K	1.20	0.80	0.05	–	–	3.20
L	3.40	4.00	5.50	5.00	6.50	5.20
M	2.50	2.40	3.00	3.20	2.50	3.50
N	9.80	6.70	7.10	6.70	6.90	8.80
O	2.50	7.50	5.40	8.70	9.80	4.10
P	0.80	1.90	3.00	2.50	3.00	1.70
Q	0.02	0.10	1.40	0.90	0.50	0.01
R	7.00	6.00	6.60	6.90	6.40	8.30
S	7.30	6.30	7.90	8.00	5.00	6.30
T	6.20	9.10	7.20	4.60	5.60	8.70
U	4.40	2.80	6.30	3.90	3.00	1.80
V	0.70	1.00	1.60	0.90	2.10	2.40
W	1.90	2.40	0.10	0.02	–	0.03
X	0.03	0.20	0.40	0.20	–	0.10
Y	0.04	2.00	0.30	0.90	–	0.60
Z	1.10	0.10	0.10	0.50	0.50	0.02

Now count the relative frequencies of all characters in a sufficiently large passage of the ciphertext and compare with the table. This will reveal quite a few unique plaintext letters. For the rest, which is not so unique, you have to puzzle a bit, which of the possibilities results in a meaningful plaintext. More problems, however, are caused by digits used in the text, for which there are of course no statistical predictions.

1.4.2 Monoalphabetic Ciphers

Instead of choosing self-discovered characters, one can just as well permute the alphabet itself, i.e. one always encrypts each plaintext letter by the same ciphertext letter. This is called a **monoalphabetic cipher**. Here is an example of such a permutation π of letters to be kept secret:

	A	B	C	D	E	F	G	H	I	J	K	L	M	N	O	P	Q	R	S	T	U	V	W	X	Y	Z
π	U	F	L	P	W	D	R	A	S	J	M	C	O	N	Q	Y	B	V	T	E	X	H	Z	K	G	I

Decryption is done with the reverse permutation π^{-1}. The ciphertext looks visually much less strange than in the Illuminati cipher, but the cipher itself has exactly the same cryptanalytic effect. The affine cipher and thus also the Caesar cipher are simple special cases of a monoalphabetic cipher.

1.5 Vigenère Cipher

1.5.1 Polyalphabetic Ciphers

We now make another attempt to generalize the Caesar or shift cipher. We now no longer shift each letter in the plaintext by the same number i of digits. Rather, we allow for different i, but the pattern should repeat after a certain period d. For example, i might cycle through 0, 21, and 4, and then the whole thing starts over with period d = 3. Figure 1.8 visualizes the procedure.

In practice, of course, the period is much larger. Therefore, it has become common practice to specify the corresponding letters in the alphabet as a so-called **keyword** instead of the sequence of numbers, in our example AVE for the values 0, 21 and 4. Thus, "A" stands for the digit "0", "V" for the "21" and "E" for the "4". The keyword must be transmitted secretly from the sender to the receiver, because it is needed to decrypt the message. Therefore, text passages from literature are often used, which do not have to be transmitted as text, but in a simpler and shorter form as a quotation, such as "Faust I, verse 512-13".

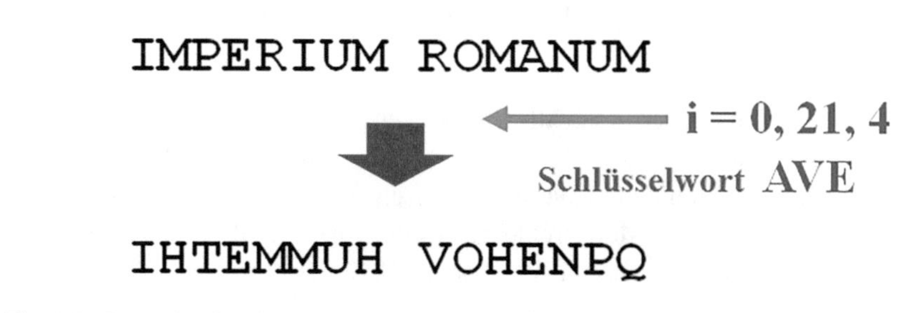

Fig. 1.8 Example of a Vigenère cipher with keyword

Our example is a **Vigenère cipher**. The diplomat **Blaise de Vigenère** (1523–1596) published it in 1586, but according to sources, the cipher was already in use since the beginning of the sixteenth century. The Vigenère cipher is a **polyalphabetic cipher**, because depending on its position in the text, the same plaintext character can be encoded by different ciphertext characters, and the same ciphertext character can stand for different plaintext characters.

1.5.2 Vigenère Tableau

Historically, the Vigenère cipher was described in a different way, namely via the so-called **Vigenère tableau**, which is shown in Table 1.3.

In the header line, it contains all 26 letters for which the cipher assignment must be determined. In the left margin column, each line is marked consecutively with the 26

Table 1.3 Vigenère tableau

	A	B	C	D	E	F	G	H	I	J	K	L	M	N	O	P	Q	R	S	T	U	V	W	X	Y	Z
A	A	B	C	D	E	F	G	H	I	J	K	L	M	N	O	P	Q	R	S	T	U	V	W	X	Y	Z
B	B	C	D	E	F	G	H	I	J	K	L	M	N	O	P	Q	R	S	T	U	V	W	X	Y	Z	A
C	C	D	E	F	G	H	I	J	K	L	M	N	O	P	Q	R	S	T	U	V	W	X	Y	Z	A	B
D	D	E	F	G	H	I	J	K	L	M	N	O	P	Q	R	S	T	U	V	W	X	Y	Z	A	B	C
E	E	F	G	H	I	J	K	L	M	N	O	P	Q	R	S	T	U	V	W	X	Y	Z	A	B	C	D
F	F	G	H	I	J	K	L	M	N	O	P	Q	R	S	T	U	V	W	X	Y	Z	A	B	C	D	E
G	G	H	I	J	K	L	M	N	O	P	Q	R	S	T	U	V	W	X	Y	Z	A	B	C	D	E	F
H	H	I	J	K	L	M	N	O	P	Q	R	S	T	U	V	W	X	Y	Z	A	B	C	D	E	F	G
I	I	J	K	L	M	N	O	P	Q	R	S	T	U	V	W	X	Y	Z	A	B	C	D	E	F	G	H
J	J	K	L	M	N	O	P	Q	R	S	T	U	V	W	X	Y	Z	A	B	C	D	E	F	G	H	I
K	K	L	M	N	O	P	Q	R	S	T	U	V	W	X	Y	Z	A	B	C	D	E	F	G	H	I	J
L	L	M	N	O	P	Q	R	S	T	U	V	W	X	Y	Z	A	B	C	D	E	F	G	H	I	J	K
M	M	N	O	P	Q	R	S	T	U	V	W	X	Y	Z	A	B	C	D	E	F	G	H	I	J	K	L
N	N	O	P	Q	R	S	T	U	V	W	X	Y	Z	A	B	C	D	E	F	G	H	I	J	K	L	M
O	O	P	Q	R	S	T	U	V	W	X	Y	Z	A	B	C	D	E	F	G	H	I	J	K	L	M	N
P	P	Q	R	S	T	U	V	W	X	Y	Z	A	B	C	D	E	F	G	H	I	J	K	L	M	N	O
Q	Q	R	S	T	U	V	W	X	Y	Z	A	B	C	D	E	F	G	H	I	J	K	L	M	N	O	P
R	R	S	T	U	V	W	X	Y	Z	A	B	C	D	E	F	G	H	I	J	K	L	M	N	O	P	Q
S	S	T	U	V	W	X	Y	Z	A	B	C	D	E	F	G	H	I	J	K	L	M	N	O	P	Q	R
T	T	U	V	W	X	Y	Z	A	B	C	D	E	F	G	H	I	J	K	L	M	N	O	P	Q	R	S
U	U	V	W	X	Y	Z	A	B	C	D	E	F	G	H	I	J	K	L	M	N	O	P	Q	R	S	T
V	V	W	X	Y	Z	A	B	C	D	E	F	G	H	I	J	K	L	M	N	O	P	Q	R	S	T	U
W	W	X	Y	Z	A	B	C	D	E	F	G	H	I	J	K	L	M	N	O	P	Q	R	S	T	U	V
X	X	Y	Z	A	B	C	D	E	F	G	H	I	J	K	L	M	N	O	P	Q	R	S	T	U	V	W
Y	Y	Z	A	B	C	D	E	F	G	H	I	J	K	L	M	N	O	P	Q	R	S	T	U	V	W	X
Z	Z	A	B	C	D	E	F	G	H	I	J	K	L	M	N	O	P	Q	R	S	T	U	V	W	X	Y

letters. The lines within the tableau also contain the entire alphabet, but always shifted cyclically to the left by one position.

Now the Vigenère keyword comes into play. Let's take the keyword `PAUSE` with the period d = 5 as an example instead of `AVE`. Then the first letter in the plaintext to be encoded is encoded according to line "P" in the tableau, the second according to line "A", then according to the lines "U", "S" and "E". At the sixth letter in the plaintext, the whole thing starts again from the beginning. To decode, the receiver must know the keyword. The receiver then deciphers with the keyword as well, inverting from the corresponding line of the Vigenère tableau back to the header line. Here is a simple example of a Vigenère cipher.

Plain text	HIERSTEHICHNUNICHARMERTOR
Key	PAUSEPAUSEPAUSEPAUSEPAUSE
Box number text	WIYJWIEBAGWNOFMRHUJQTRNGV

1.5.3 Smoothing of Statistical Frequencies

The main advantage of the Vigenère cipher is that it smoothes the statistical frequency of letters in natural languages. Let us again take the keyword `PAUSE` and as an example the ciphertext letter I. We take from the Vigenère tableau that I may have arisen from the following plaintext letters: T, I, O, Q, and E. Here are the rounded statistical frequencies of these letters in the German language:

$$T: 6\% \qquad I: 8\% \qquad O: 2\% \qquad Q: 0\% \qquad E: 17\%$$

So we see that I can arise from the most frequent letter E, but also from the almost non-existent Q, from the relatively frequent I and from the quite rare O, as well as from the averagely frequent T. So with our statistical approach alone, we no longer get to the bottom of the Vigenère cipher. For this we need a new idea.

1.6 Kasiski and Friedman Attack

1.6.1 Kasiski Attack

So let's do some cryptanalysis again and try to crack the Vigenère cipher. As we will see in a moment, this is not too difficult if the keyword is relatively short and thus the period d is relatively small compared to the length of the ciphertext. The attack is then done in two steps.

- First we determine the length of the keyword, i.e. the period d. For this we will learn the Kasiski attack in a moment.

- Then one determines the keyword itself. If the period d is known, then it is only a matter of d different shift ciphers, because the positions 1, d + 1, 2 • d + 1,... are encrypted by identical shift ciphers, as are the positions 2, d + 2, 2 • d + 2,... and so on. As is well known, these can easily be cracked individually using statistical methods (Sect. 1.3).

The **Kasiski attack** goes back to **Friedrich Wilhelm Kasiski** (1805–1881), who published the method in 1863. It is based on the following idea: If certain character sequences occur frequently in the plaintext, they also become identical ciphertext sequences if their spacing is a multiple of the period d. Such repetitions of strings can of course also occur randomly in the ciphertext, but this is much less often the case. Thus, in the Kasiski attack, the ciphertext is examined for repetitions of strings of at least three characters and the respective distances are determined. This gives clear indications of the period d. As a rule, only a few possibilities remain, which are then examined more closely. This is best seen in an example [Hau1], for example in the following ciphertext:

```
GTHFYW FWJJTB NHZGOY FKUNVS NQOZNF GNQTJI FGJYVF IRGPPV
QHJBWJ LKTXMU JJAKJK FMYWPU ZCLNFD HZVFMX SIFIPS POJGWT
FILDKT XMGFFM KJCOUT WUJNHZ KTOFDU JCSRWF IYKEIG ZWUTZQ
FFJZXY KMOSSV VZWDKD CFMEIA ZWTLGF JVFSYW UHDGLG JIJAPG
VHZGTQ JJHBCW WPHZSM GHZSVM BILWPL GFJVFS YSWGUZ YGIJJJ
KHFJAV APJAPF MGWMBI SLGONU JCDCJR WVIYWT TPHPGO JMFGTZ
NFGTTS LCYPSV UFHFFV JFEMDF OWSEIO JF
```

Table 1.4 lists some of the repeating strings with their positions and the respective distance.

Since the long string LGFJVFSY is very unlikely to repeat randomly, we can assume that the period is a divisor of 50. Since the number 5 divides the occurring intervals in nine

Table 1.4 Kasiski attack

String	Position 1	Position 2	Distance
NHZ	13	118	$105 = 3 \cdot 5 \cdot 7$
HZG	14	194	$180 = 2^2 \cdot 3^2 \cdot 5$
ZNFG	28	288	$260 = 2^2 \cdot 5 \cdot 13$
KTXM	56	101	$45 = 3^2 \cdot 5$
UJC	126	264	$138 = 2 \cdot 3 \cdot 23$
LGFJVFSY	172	222	$50 = 2 \cdot 5^2$
JAP	189	249	$60 = 2^2 \cdot 3 \cdot 5$
HZS	207	212	5
MBI	216	256	$40 = 2^3 \cdot 5$
FGT	280	285	5

cases, but 2 only in six cases and 25 only in one case at all, it is plausible to assume that the period d = 5. The string `UJC` would then have been the only one to repeat at random. With d = 5 one now goes into the statistical analysis. If this does not lead to success, one would still try the second most plausible possibility d = 10. So the Kasiski attack also needs some luck to find some strings as long as possible that repeat.

1.6.2 Friedman Coincidence Index

The **Friedman attack**, published in 1918 by **William Frederic Friedman** (1891–1969), takes a more systematic approach, but also requires a bit more mathematics. If you are put off by this, you can just "skim" this point or even skip it.

The attack uses the so-called **Friedman coincidence index** I, which indicates the probability that in a sufficiently long string of characters at two randomly selected positions of the sequence are the same characters. To calculate this index at least approximately, we imagine a text consisting of m capital letters, where each of the 26 letters z = 0,…, 25 may occur exactly m_z -times. Then the relative frequency of choosing the letter z twice given a random choice of two letters is given by $m_z \bullet (m_z - 1)/(m \bullet (m - 1))$. Overall, then, the relative frequency of choosing any letter twice is equal to $(m_0 \bullet (m_0 - 1) + \ldots + m_{25} \bullet (m_{25} - 1))/(m \bullet (m - 1))$, giving us an approximate formula for I.

We determine I in another way, but now the characteristic of the chosen language enters. Let p_z be the probability for the occurrence of the letter z in this language. For some languages we have already listed these probabilities in Table 1.2. The probability I of choosing two identical letters from an arbitrary but sufficiently long text is thus $I = p_0^2 + p_1^2 + \ldots + p_{25}^2$. For the German language, for example, this gives the value $I_D = 0.0762$ and for the English language $I_E = 0.0611$. For a fictitious language in which all letters have the same probability, $I_{gW} = p_0^2 + p_1^2 + \ldots + p_{25}^2 = 26 \cdot (1/26)^2 = 1/26 = 0.0385$.

1.6.3 Friedman Attack

To determine the period d of a Vigenère cipher using the Friedman attack, we think of the ciphertext with its m letters read row by row into a table with d columns. Moreover, we now also concretely assume that the plaintext comes from the German language. Then we can state the following two facts:

- Each column of this table was encoded using a shift cipher. In particular, their letter distribution corresponds to that of the German language.
- On the other hand, the Vigenère cipher smooths the overall letter frequency. Therefore, we can approximately assume that all letters occur with equal probability in the entire table.

Thus, if we select two letters of the ciphertext in the same column of the table, the probability of drawing the same letter twice is approximately $I_D = 0.0762$, but if the two letters are from different columns, this probability is approximately $I_{gW} = 0.0385$. Now, for the number g of pairs of letters from the same column, $g = m \bullet (m/d - 1)/2$, and for the number v of pairs of letters from different columns, $v = m \bullet (m - m/d)/2$. Moreover, $m \bullet (m - 1)/2$ is the number of pairs of letters in the entire table and hence in the entire ciphertext. Consequently, I is approximated given by the equation $I = (g \bullet I_D + v \bullet I_{gW})/(m \bullet (m - 1)/2) = (0.0762 \bullet m \bullet (m/d - 1)/2 + 0.0385 \bullet m \bullet (m - m/d)/2)/(m \bullet (m - 1)/2)$, which resolves to $d = 0.0377 \bullet m/((m - 1) \bullet I - 0.0385 \bullet m + 0.0762)$. Inserting here the calculation formula $I = (m_0 \bullet (m_0 - 1) + \ldots + m_{25} \bullet (m_{25} - 1))/(m \bullet (m - 1))$ gives an at least approximate formula for determining the period d.

1.7 Enigma Machine

1.7.1 Structure of the Enigma Machine

As a conclusion of the chapter about historical ciphers, we will now report about a cipher machine which was used by the German Wehrmacht during World War II: the **Enigma machine** (gr. ainigma, engl. riddle). The inventor is **Arthur Scherbius** (1878–1929), whose first patent dates back to 1918.

In the Enigma machine, the plaintext is entered via a keyboard. If one presses a letter key, electric current flows by means of a battery in the Enigma over an ingenious arrangement of circuits and finally lets light up an indicator lamp in the lamp field, which indicates the coding of the pressed letter. Typically, in this circuitry, the electrical signal is first fed to a plugboard. This has 26 contacts, one for each letter of the alphabet. Of these 26 contacts, ten pairs are selected to be wired together. Figure 1.9 shows the schematic of the Enigma plugboard with ten exemplary wirings visualized in red. So the signal from the keyboard is possibly redirected to another letter on the plugboard. Non-wired letters remain unchanged.

The signal is then applied to a roller set consisting of three rollers. Each roller has 26 input and output contacts, which are interconnected in pairs within the roller. As shown schematically in Fig. 1.10, the signal arriving on the left for a letter is first passed along the

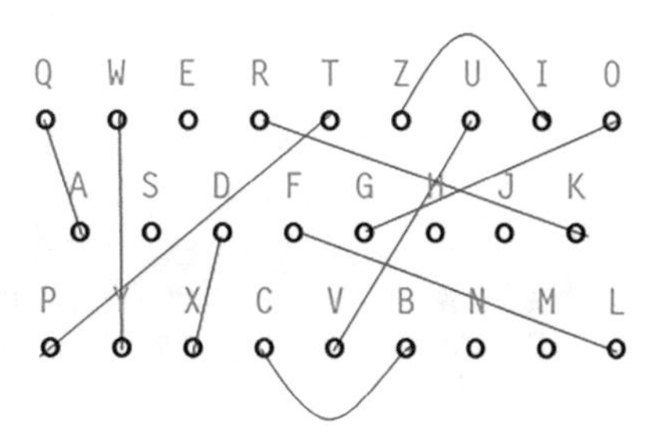

Fig. 1.9 Schematic of the plugboard of the Enigma machine

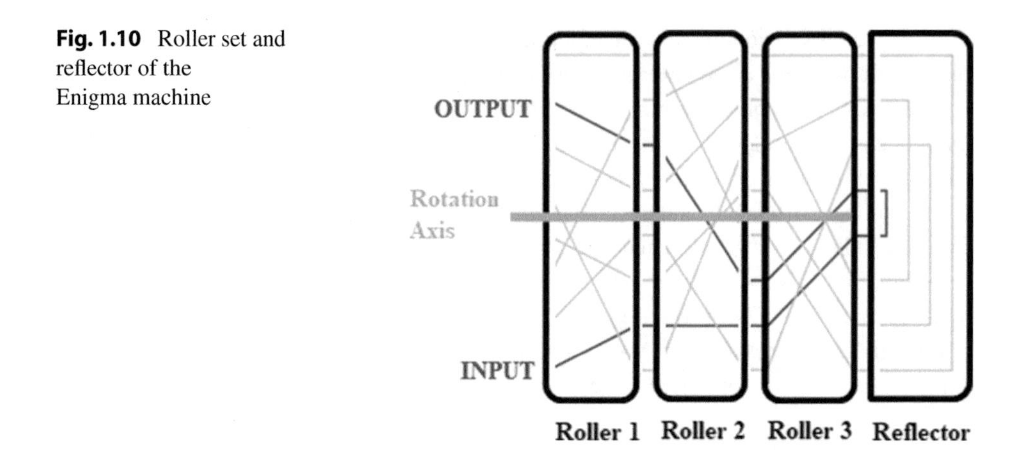

Fig. 1.10 Roller set and reflector of the Enigma machine

red path through the three rollers and then meets a reflector, which in turn has 26 contacts connected in pairs. The reflector passes the signal back through the three rollers. Each roller consists of two parts; the core with the fixed wiring for substitution and the ring. The ring position determines the offset between the internal wiring of the rollers and the letter at which the carryover to the next roller occurs. After exiting the roller set, the signal is passed over the plugboard again and then ultimately displayed in the lamp field as a letter.

While the reflector is immobile, the three rollers, driven by a mechanical coupling, rotate as follows after each input of a letter: The left "fast" roller starts and rotates one position after each letter is entered, so that it returns to its original position after 26 rotational steps. After that, the "middle" roller rotates by one position, and then the first one again completes a full rotation. After 26 rotation steps of the "middle" roller, the right "slow" roller starts to rotate by one position, and this continues until the end of the text. Due to its rotation mechanism, the Enigma machine is also a polyalphabetic cipher.

1.7.2 Configuration of the Enigma Machine

Basically, the five rollers I to V and the three reflectors A to C were available for the Enigma machine [WPEnM]:

	A	**B**	**C**	**D**	**E**	**F**	**G**	**H**	**I**	**J**	**K**	**L**	**M**	**N**	**O**	**P**	**Q**	**R**	**S**	**T**	**U**	**V**	**W**	**X**	**Y**	**Z**
I	E	K	M	F	L	G	D	Q	V	Z	N	T	O	W	Y	H	X	U	S	P	A	I	B	R	C	J
II	A	J	D	K	S	I	R	U	X	B	L	H	W	T	M	C	Q	G	Z	N	P	Y	F	V	O	E
III	B	D	F	H	J	L	C	P	R	T	X	V	Z	N	Y	E	I	W	G	A	K	M	U	S	Q	O
IV	E	S	O	V	P	Z	J	A	Y	Q	U	I	R	H	X	L	N	F	T	G	K	D	C	M	W	B
V	V	Z	B	R	G	I	T	Y	U	P	S	D	N	H	L	X	A	W	M	J	Q	O	F	E	C	K

A	AE	BJ	CM	DZ	FL	GY	HX	IV	KW	NR	OQ	PU	ST
B	AY	BR	CU	DH	EQ	FS	GL	IP	JX	KN	MO	TZ	VW
C	AF	BV	CP	DJ	EI	GO	HY	KR	LZ	MX	NW	QT	SU

This results in the following degrees of freedom in the configuration of the Enigma machine:

- Selection of reflector: 3 possibilities
- Selection and position of the three rollers: $5 \cdot 4 \cdot 3$ possibilities
- Determination of the initial position of the three rollers: 26^3 possibilities
- Determination of ring positions: 26^3 possibilities
- Definition of the ten plug connections: $(26 \cdot 25 \cdot 24 \cdot \ldots \cdot 8 \cdot 7)/2^{10}$ possibilities

1.7.3 Deciphering and Security with Enigma

The trick of the machine is that by mirroring the incoming signal at the reflector every substitution of a letter is involutory, i.e., if the letter X is enciphered in Y, the letter Y would also have been substituted in X at this text passage. Therefore, if the receiver of messages was secretly informed of the chosen configuration, he could decode the received message with exactly the same configured Enigma machine.

The configuration options of Enigma were enormous for that time. Nevertheless, many of them do not contribute much to Enigma's security. The plugboard, for example, provides nothing more than a simple monoalphabetic substitution. Also the facts that Enigma is involutory and fixed-point free, i.e. can never substitute a letter by itself, provide points of attack for cryptanalysis. The Enigma encryption was then also cracked by the British through the group around **Alan Turing**.

2 Symmetric Ciphers

2.1 Keys and Attack Strategies

It is now time to understand our approach to ciphering in a slightly more conceptual way.

2.1.1 Algorithm and Key

Every encryption method consists of an **encryption algorithm** together with one or more parameters, the **encryption key**, which can be used to assign a ciphertext one-to-one to any plaintext. The cipher key thus determines the characteristics of the algorithm.

In the case of the shift cipher, the algorithm describes the "shifting" or, mathematically speaking, the "addition modulo 26". The key corresponds to the shift value i, i.e. in the case of the Caesar cipher $i = 3$. In the case of the affine cipher, one even needs a key pair, namely (i, j). In the case of the Vigenère cipher, the Vigenère tableau describes the algorithm and the keyword, or the text passage used as the key, already has a whole set of individual parameters, namely the letters of the keyword or text.

The requirement of the ciphertext being one-to-one means that for each ciphertext its corresponding plaintext can be uniquely determined. The decryption of the ciphertext is therefore carried out with a **decryption algorithm** belonging to the encryption method, which in turn depends on one or more parameters, the **decryption key**.

The decipherment algorithm for the shift cipher is also a shift cipher, and the decipherment key is -i. The decipherment algorithm for the affine cipher is again an affine cipher, and the decipherment key consists of the key pair (j^{-1} (mod 26), $-j^{-1} \bullet i$ (mod 26)). In the decipherment algorithm for the Vigenère cipher, one inversely infers from the rows of the Vigenère tableau to the header row, using the same keyword or keytext that is used in the

O. Manz, *Encrypt, Sign, Attack*, Mathematics Study Resources,
https://doi.org/10.1007/978-3-662-66015-7_2

cipher. The decipherment algorithm of the Enigma machine works with an exactly identically configured Enigma machine.

2.1.2 Symmetric and Asymmetric Ciphers

In our examples, the decryption key could easily be determined from the encryption key, in some cases both were even the same. In this case we speak of **symmetric ciphers**. With symmetric ciphers, knowledge of the cipher key is sufficient for decryption; in short, one simply speaks of the **key** of a symmetric cipher.

If, on the other hand, the decryption key cannot be calculated from the encryption key, or can only be calculated with an extremely large amount of effort and therefore in an unrealistically long time, we speak of **asymmetric ciphers**. In this chapter we will exclusively deal with symmetric ciphers and only deal with asymmetric ciphers in the next chapter.

2.1.3 Kerckhoff's Principle

A symmetric encryption method therefore always contains an encryption and decryption algorithm, which in turn depend on parameters, the key. Communication partners must therefore first agree in principle on the algorithms to be used. Experience shows that this cannot be kept completely secret, especially not in the long run, since one does not want to constantly change the algorithms. If one therefore assumes that the algorithms are in principle known to a potential attacker, then the entire security of the procedure ultimately depends on the secrecy of the key. In cryptography, the **Kerckhoffs principle** named after **Auguste Kerckhoffs** (1835–1903) is thus always required: The security of a symmetric encryption method may only depend on the secrecy of the key, but not on the secrecy of the algorithms.

2.1.4 Military Secrets

Despite Keckhoff's principle, the algorithms of many encryption processes are still kept secret, especially in the military and intelligence sectors. An example is the satellite navigation system GPS of the USA. A satellite navigation system is based on several satellites that constantly broadcast their current position and the exact time using radio signals. Special receivers can then calculate their own position from the signal propagation times of four satellites. There are several systems worldwide, in particular **GPS** (Global Positioning System) of the USA, **Galileo** (EU), **GLONASS** (Russia) and **Beidou** (China). GPS dates back to the late 1980s and was originally developed for navigation by the US Navy (NAVSTAR GPS). Today, however, it is at least partially available for civilian use and is the de facto standard on many roads. On the so-called L1 carrier frequency of 1575.42 MHz, the **C/A code** (Coarse/Acquisition) is transmitted as the basis for civilian use. The non-public **P/Y code** (Precision/Encrypted) for precise military positioning is transmitted separately on top of this. To protect against a possible enemy, the P-code is encrypted into a Y-code. The procedure is kept secret by the military as a whole.

2.1.5 Attack Strategies

Based on the Kerckhoffs principle, the key of a symmetric encryption method must be communicated to the authorized recipient in a secure way. However, there are several reasons why the entire message is not transmitted in this secure way:

- The message is rather long, the key relatively short.
- The time of the handing over of the key is freely selectable.
- Multiple messages can be encrypted and decrypted with the same key.

To crack a cipher, one therefore probably comes up with the obvious idea of obtaining the secretly exchanged key in some way. But on the one hand, the handover of the key was also particularly well secured for historical ciphers. On the other hand, there are now modern methods of key exchange, which we will get to know in the next chapter, that make such an attempt hopeless from the outset.

So you might try to brutally check all possible keys one after the other. This is called a **brute-force attack**. If necessary, the sequence can be selected according to probabilities known from experience. Even with modern encryption methods, this method is always useful if the key space is not large enough. In this case, networked computers may be able to calculate all possibilities in a reasonable amount of time.

The most elementary variant is that an attacker listens to the entire ciphertext or at least large parts of it and tries to use it to find the key or at least to deduce the corresponding plaintext. This attack is called a **ciphertext-only attack**. The statistical cryptanalysis of shift ciphers and more general monoalphabetic ciphers as well as the pattern recognition of the Kasiski attack are typical examples.

The **known plaintext attack** has a greater chance of determining the key. The attacker listens to the ciphertext, but also knows parts of the plaintext or at least assumes to know them. For example, Enigma could be cracked with the knowledge that event messages always started with the place and date and that the daily weather report was routinely sent. Another example is the attack on the old encryption method WEP of the WLAN (Sect. 3.2), which exploits the fact that the encrypted header data of the WLAN protocol are predictable.

The **chosen plaintext attack** is even more powerful. Here, the attacker is able to have plaintext passages of his choice encrypted to a certain extent. To do this, the attacker must, for example, be able to foist the messages to be encrypted on the victim in such a way that the victim is not aware of this. Or he has at least temporary access to the encryption device, for example through a break-in or theft, without the current key being directly readable (e.g. on a smartphone). In this way, the plaintext can be varied and the resulting changes in the ciphertext can be analyzed.

Finally, there is also the **chosen ciphertext attack**, whereby the attacker temporarily even has the possibility of having ciphertexts of his choice decrypted to a certain extent. This is the case, for example, if the encryption device is also used for decryption and the attacker has at least temporary access to the device, for example by stealing it. This attack is often fatal for the security of the procedure.

Cryptanalysis is not only carried out by attackers with the aim of cracking a cipher procedure and thus eavesdropping on the secret information, but also by cryptographers in order to prove or quantify the security of the procedure.

2.2 Vernam Cipher and Pseudo-Randomness

2.2.1 Vernam Cipher

Gilbert Vernam (1890–1960) patented the following method surprisingly already in 1918. Vernam interpreted the plaintext as a binary string consisting of the bits 0 and 1 and therefore already worked digitally (Sect. 1.2). The key of his **Vernam cipher** is a randomly generated bit string, which is as long as the plaintext and is added bit by bit to the plaintext string $\oplus$ to encrypt it. One advantage of the method is that to decrypt it, one simply has to add the same bit string back up $\oplus$.

Plaintext	11000101...
Key	$\oplus$ 01101100...
Ciphertext	10101001...

But what does a random sequence of bits mean? The decisive factor is how the sequence was generated, namely each bit as an independent fair coin toss, with probability 1/2 for both the 0 and the 1. The Vernam cipher is made particularly secure if such a bit sequence is used only once for encryption. This is called a **one-time pad**.

2.2.2 Shift Registers

So much for the theory of the Vernam cipher. But now the cat bites its own tail, because to encrypt plaintext, you need a randomly generated bit string of the same length as a key, which you have to communicate secretly to the recipient beforehand so that he can decrypt it. How can this be done in a practicable way? In order to get out of this dilemma, a method has been devised in which a much shorter key can be exchanged and in which the receiver is nevertheless able to generate the bit string used by the sender himself. This method is based on digital switching elements, so-called **linear feedback shift registers**, the function of which we will now illustrate using the example shown in Fig. 2.1 [Man, Beu].

A shift register has m cells, each with one bit $z_1, \ldots, z_m$ as cell content. The sender secretly informs the receiver of the initialization of the cell contents $z_1 = i_1$ to $z_m = i_m$. In our example, $i_1 \ldots i_8 = 01100101$. In addition, the sender secretly informs the receiver whether to interconnect after a cell ($v_j = 1$) or not ($v_j = 0$). In our example, this means $v_1 \ldots v_8 = 01010011$. With each clock pulse of the switching element, the contents of the cells are shifted one position to the right and the last bit is output on the right. Via the so-called feedback equation $z_1 \bullet v_1 + \ldots + z_m \bullet v_m$, the first cell is simultaneously filled again. In our example, this means that for the first clock pulse $z_1 \bullet v_1 + \ldots + z_8 \bullet v_8 = i_1 \bullet v_1 + \ldots + i_8 \bullet v_8 = 0 \bullet 0 + 1 \bullet 1 + 1 \bullet 0 + 0 \bullet 1 + 0 \bullet 0 + 1 \bullet 0 + 0 \bullet 1 + 1 \bullet 1 = 1 + 1 = 0$, where the addition and multiplication of bits is meant here (Sect. 1.2). So 1 is output on the right and the first cell is filled with 0. In the second clock pulse, the new cell contents z_1 to z_m are used and the procedure is repeated. The whole thing can therefore be continued as often as desired. In this way, the sender and receiver can exchange the comparatively short key consisting of the $2 \bullet m$ bits $i_1 \ldots i_m$ $v_1 \ldots v_m$ for a shift register of length m and thus generate the same, initially rather random-looking bit sequence. This can then be added bitwise $\oplus$ to a plaintext of arbitrary length as a **shift register cipher**, as in the Vernam cipher.

2.2.3 Pseudo-Randomness and Cryptanalysis

But sequences generated by linear feedback shift registers are no real random sequences, because the next bit is always determined by the current contents of the m cells. The sequence is even periodic with a period length at most $2^m - 1$. This follows simply from the fact, that the m cells can take at most 2^m different values z_1 to z_m, and if all $z_1 = \ldots = z_m = 0$, only one sequence consisting of all 0 is generated.

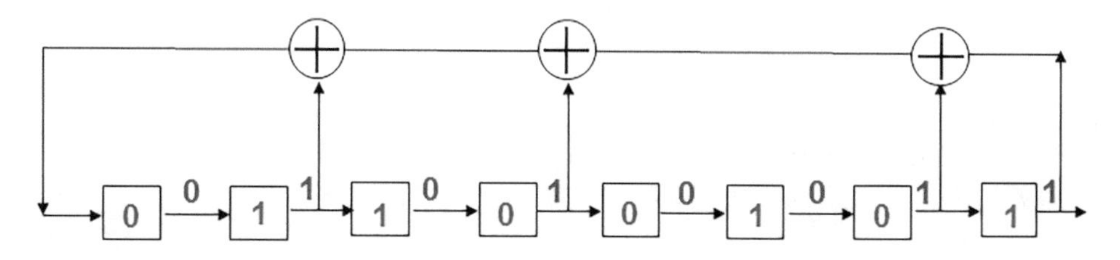

Fig. 2.1 Example of a linear feedback shift register with initialization (blue) and interconnection (red)

Thus with linear feedback shift registers only **pseudo-random sequences** can be generated. It is true that there are criteria for a shift register to have maximum period $2^m - 1$, and in practice it is advisable to only use such in shift register ciphers in the first place. However, if an attacker can obtain only $2 \bullet m$ consecutive bits of the pseudo-random sequence generated by the shift register, he knows the entire formation law and can decipher at his leisure from then on. All he has to do is solve a system of equations with m equations and m unknowns. We consider this with a small example and imagine that an attacker has identified a sequence of $2 \bullet m = 6$ bits as part of a pseudo-random sequence generated by a shift register:

Shift Register	→	… 1 0 1 1 0 0 …

Then surely the following three feedback equations hold:

$$1 = 1 \bullet v_1 + 0 \bullet v_2 + 0 \bullet v_3 \rightarrow v_1 = 1$$
$$0 = 1 \bullet v_1 + 1 \bullet v_2 + 0 \bullet v_3 \rightarrow v_2 = 1$$
$$1 = 0 \bullet v_1 + 1 \bullet v_2 + 1 \bullet v_3 \rightarrow v_3 = 0$$

These immediately provide the attacker with the shift register he is looking for with current initialization, as shown in Fig. 2.2.

Shift register ciphers that use only a single linear feedback shift register are therefore completely unsuitable for cryptographic practice. In order to be able to use their technical advantages nevertheless, several shift registers are sometimes concatenated (Sect. 2.3).

2.3 GSM Mobile Communications

2.3.1 The GSM Mobile Communications Standard

GSM (Global System for Mobile Communications) is the standard for digital mobile communications networks of the so-called 2. generation **(2G)** as successor of the analogue networks of the first generation. It was primarily designed for telephony and short messages (SMS Short Messages). GSM was introduced in Germany in 1992 and is still used today by many mobile phone customers worldwide.

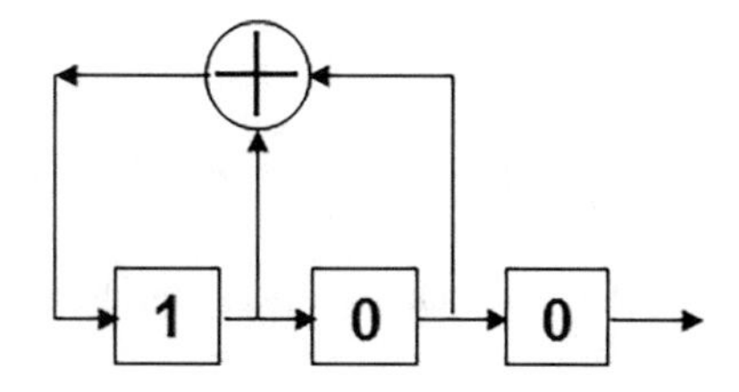

Fig. 2.2 Example of a small linear feedback shift register

2.3.2 GSM Data Encryption

For data encryption, GSM uses the algorithms **A8** for key generation and **A5** for the actual encryption of the telephone call or SMS. A5 is a procedure which was initially designed in 1987 as A5/1 and in 1989 additionally in a weakened version for certain export regions as A5/2. Originally, an attempt was made to keep the algorithm secret, contrary to the Kerckhoffs principle, but this failed. In the meantime, however, A5/1 is open and standardized. The algorithm A8 is defined by the respective network operator and kept secret as far as possible.

GSM data encryption, which is visualized in Fig. 2.3, uses personalized chip cards (**ICC** Integrated Circuit Card). These so-called **SIM cards** (Subscriber Identification Module) are issued by the network operators to their customers. Each subscriber is thus assigned a 128-bit subscriber key k_i (Subscriber Authentication Key), which is stored on the SIM card on the one hand and in the mobile communications server on the other. The mobile network also sends a 128 bit long random number RAND when the subscriber logs on. The A8 algorithm uses RAND and the subscriber key k_i to generate a 64-bit key k_c on the subscriber's SIM card and in the mobile communications server. The A5 algorithm together with the key k_c ultimately performs the encryption and decryption of the calls and SMSs. (For subscribers' authentification confer Sect. 4.9.)

2.3.3 A5 Cipher of Version A5/1

The A5 algorithm of version A5/1 is a shift register cipher with three linear feedback shift registers connected in parallel. For encryption, the outputs of all three shift registers are added in binary and added to the plaintext. Figure 2.4 shows the structure.

In contrast to the shift register ciphers described so far, however, the lengths and interconnections are publicly known here, i.e. they are part of the algorithm. The same applies

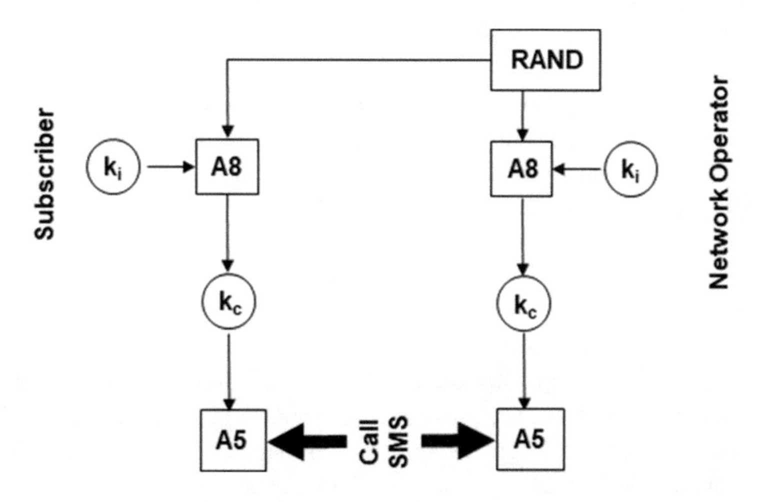

Fig. 2.3 Key generation and data encryption for GSM

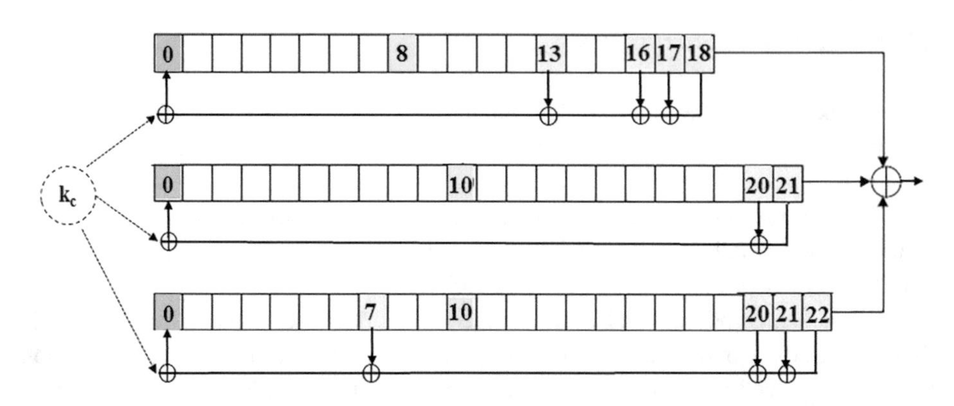

Fig. 2.4 Shift register cipher A5/1 for GSM

to the initialization of the cells. At the beginning, they all contain the value 0. Only now does the 64-bit cipher key k_c come into play. It is successively loaded into the first cell of each of the three shift registers by bitwise addition $\oplus$. In this process, the shift registers are clocked 64 times, and in each case another bit of the cipher key k_c is added to the contents of the first cells. After that, the registers are clocked several times irregularly, depending on the contents of the cells 8, 10 and 10 highlighted in yellow. The output bits expire unused, and only then does the actual encryption begin by binary addition to the plaintext [Sto].

The method A5/1 and especially the similar but weaker version A5/2 are considered insecure, the encryption cannot provide significant security against serious attacks [Sto]. But at least it prevents simple eavesdropping. The successor versions A5/3 and A5/4, which are considered secure, differ fundamentally from A5/1 (Sect. 2.7).

2.4 Feistel Cipher

2.4.1 Stream Ciphers and Block Ciphers

Stream ciphers are encryption methods in which the sequence of plaintext characters is encrypted one after the other and (pseudo-)randomly varying in each step. If, on the other hand, the plaintext is divided into blocks of fixed length, which are all encrypted separately, and the encryption method is the same for each block, this is known as **block ciphers**. Thus, in order to design secure ciphers, one either invests in the costly generation of the key or in complex encryption methods on blocks of suitably large length, where the key can be chosen more simply. An advantage of stream ciphers compared to block ciphers is that one can decrypt character by character and does not always have to wait for a whole ciphertext block.

In this sense, the Vernam cipher and the shift register cipher are stream ciphers. The shift and affine ciphers are block ciphers of block length 1. The Vigenère cipher is also a block cipher, with the block length determined by its period d. Most of today's important ciphers are or are at least based (Sect. 2.6) on block ciphers.

2.4.2 Confusion and Diffusion

But what are the quality criteria that should be applied to a block cipher? **Claude Shannon** (1916–2001) formulated two rather intuitive criteria as early as 1949, but they still hold true today.

- **Confusion:** As far as possible, no relationship should be recognizable between plaintext and ciphertext that could be exploited for an attack. This applies in particular to the statistical distribution of the characters in the plaintext and ciphertext.
- **Diffusion:** All characters of the plaintext and of the key should influence as many characters of the ciphertext as possible.

2.4.3 Construction Principle of a Feistel Cipher

We now come to the prototype of modern block ciphers par excellence, the **Feistel cipher**, which goes back to **Horst Feistel** (1915–1990). In 1973, under the project name **LUCIFER,** he developed an encryption method that can be regarded as the forerunner of the DES (Data Encryption Standard) (Sect. 2.5).

However, the Feistel cipher is rather a construction principle for a block cipher, which is composed of an arbitrary number of so-called **rounds**. The plaintext $m = m_1 \ldots m_n$ is taken as a binary string and divided into blocks m_1 to m_n of even length $2 \bullet t$, where t is arbitrary. It may be necessary to suitably pad m_n in the process. Each of these blocks is now ciphered separately. So we consider a fixed such block $L_0\, R_0$ with binary strings L_0 and R_0 of length t. Let further $F(\bullet, \lambda)$ be an arbitrary function that transforms a binary string of length t into a binary string of length t and that has as parameter a binary string λ of arbitrary length. Furthermore, let k_i be a bit string of the same length as the parameter λ, the so-called **round key** for the i. round. We now want to describe the so-called **round function** of a Feistel cipher, and this is done recursively. Let $L_{i-1}\, R_{i-1}$ be the binary string of length $2 \bullet t$, which has arisen after the $(i - 1)$. round. Then the round function that computes the next binary string $L_i\, R_i$ at the i-th round of a Feistel cipher is as follows:

$$
\begin{aligned}
L_i &= R_{i-1} \\
R_i &= L_{i-1} \oplus F\left(R_{i-1}, k_i\right)
\end{aligned}
$$

In words, it means this:

- Place the right side R_{i-1} on the left side,
- apply to the right side R_{i-1} the mapping $F(\bullet, k_i)$ with the round key k_i,
- add $\oplus$ this binary string position by position to the left side L_{i-1}
- and place this sum on the right side.

It is best to look at the first two steps graphically, as shown in Fig. 2.5.

In principle, therefore, any number of rounds is possible with a Feistel cipher. At the same time, however, the number of round keys k_i and thus the size of the total key k_1, k_2, k_3… increases considerably. Therefore, a Feistel cipher always includes the basic idea of generating the individual round keys k_i conversely from a relatively short "base key". Here are the essential advantages of a Feistel cipher.

- Ciphering and deciphering is done with exactly the same algorithm, where you only have to apply the round keys in reverse order. This has the advantage for the computer implementation that the same program modules are sufficient for both. Figure 2.6 shows the last two rounds of decryption.
- However, this also means that, unlike a cipher, the function $F(\bullet, \lambda)$ need not be one-to-one, i.e., one has significantly more degrees of freedom in a concrete realization of the Feistel cipher.
- Finally, the Feistel cipher effectively operates on only half the block length t and is therefore much faster to implement.

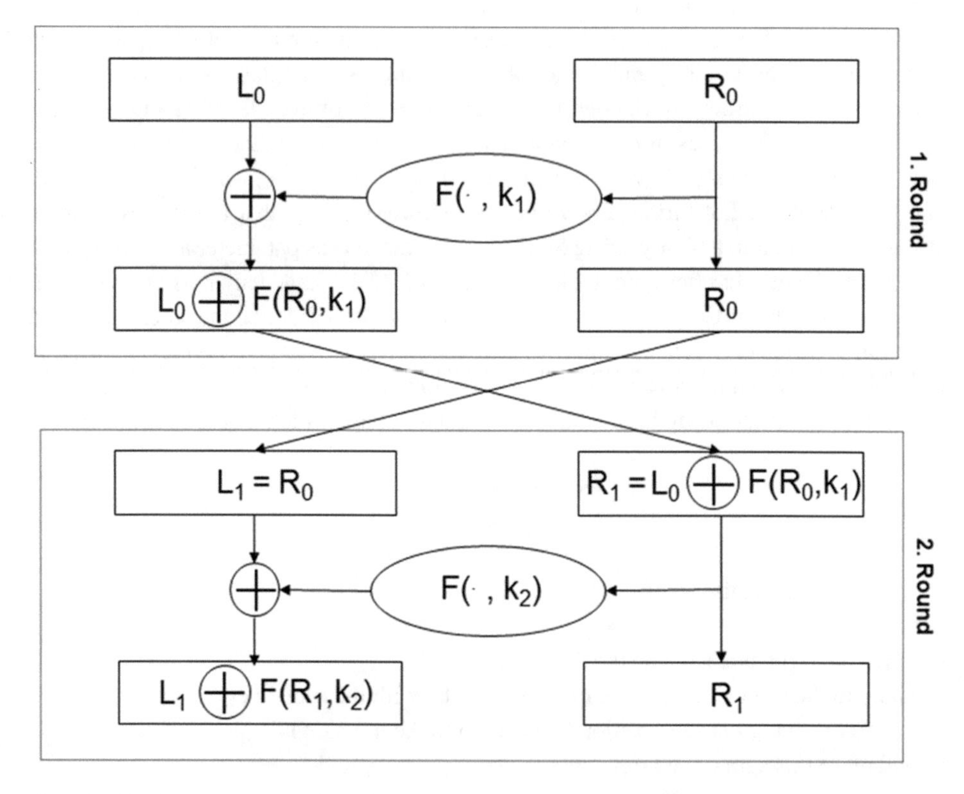

Fig. 2.5 The first two rounds of a Feistel cipher

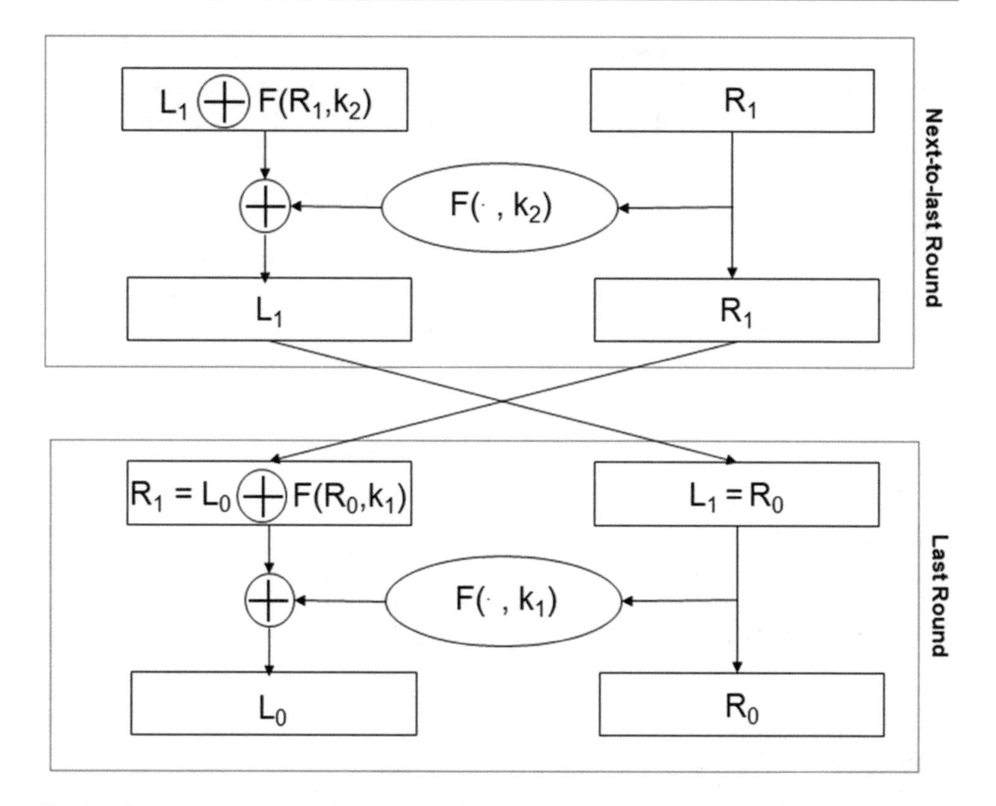

Fig. 2.6 The last two rounds of Feistel deciphering

2.5 Data Encryption Standard DES

2.5.1 From LUCIFER to DES

In 1973 and 1974, the US standardization authority **NIST** (National Institute of Standards and Technology) issued two calls for proposals for a standardized cryptographic algorithm. After none of the candidates appeared to be suitable in the first tender, **LUCIFER** from IBM remained as the only acceptable proposal in the second tender.

In the course of the assessments by the **NSA** (National Security Agency), numerous modifications were made and many adjustments were made, as will certainly become clear from the following description of the procedure. In 1977, the **DES (Data Encryption Standard)** came into force in its final form. DES was the first cryptographic algorithm ever to be standardized, with all details published. Subsequently, the standard was reviewed and extended every 5 years.

2.5.2 DES as Feistel Cipher

The **DES** is a block cipher with 64-digit binary input and output blocks. Its key is also formally 64 bits long. However, it effectively consists of eight strings with seven bits each, to which one bit each is appended for parity checking for error detection, i.e. effectively 56 bits in total. More precisely, DES is a Feistel cipher with a total of 16 rounds. Figure 2.7 gives a first overview of a DES round. The 16 rounds are preceded and followed by fixed, mutually inverse input and output permutations on 64 bits, which do not play any role for the security of DES [Buc, Hau1]. As with all Feistel ciphers, DES is decrypted using the same algorithm with the round keys in reverse order.

We now proceed to the round function of the i. round and thus ultimately to the choice of the mapping $F(\bullet, \lambda)$ in DES. Here, we first replace the parameter λ by a 48-bit round key k_i, which in turn is derived from the 56 bits of the effective DES key. We will first postpone how exactly this is done. So we first describe in the four steps

- Expansion
- Key addition
- S-Boxes
- Permutation,

as $F(\bullet, k_i)$ operates on the right 32-bit block R of a 64-bit block LR.

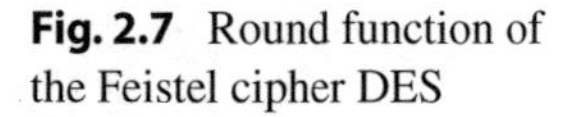

Fig. 2.7 Round function of the Feistel cipher DES

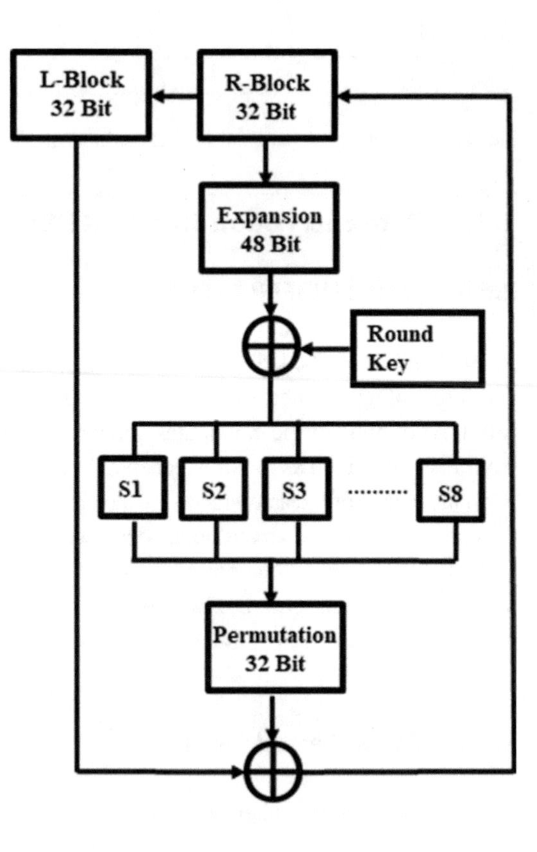

2.5.3 DES Expansion

First the 32-bit string R is enlarged to 48 bits with an **expansion** ε. To do this, the 32-bit string $a_1\, a_2 \ldots a_{32}$ is divided into eight sub-blocks of 4 bits each and each of these sub-blocks is expanded by the edge bit of the predecessor and successor sub-block to 6 bits.

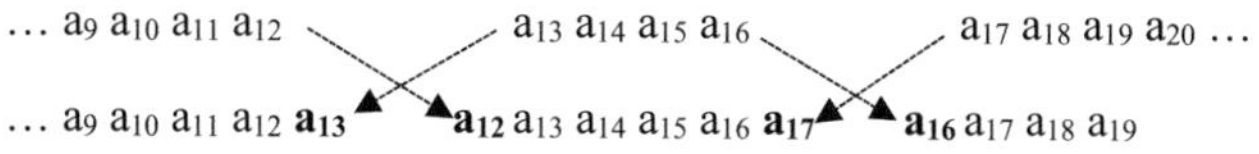

$$\ldots a_9\, a_{10}\, a_{11}\, a_{12} \quad a_{13}\, a_{14}\, a_{15}\, a_{16} \quad a_{17}\, a_{18}\, a_{19}\, a_{20} \ldots$$

$$\ldots a_9\, a_{10}\, a_{11}\, a_{12}\, \mathbf{a_{13}} \quad \mathbf{a_{12}}\, a_{13}\, a_{14}\, a_{15}\, a_{16}\, \mathbf{a_{17}} \quad \mathbf{a_{16}}\, a_{17}\, a_{18}\, a_{19}$$

The last bit of R is used at the beginning of the first block and the first bit of R is used at the end of the last block.

2.5.4 DES Key Addition

To this 48-bit string we add position by position $\oplus$ the 48-digit round key k_i. After **key addition**, we now call the bits b_j and b_j', respectively. Here, the b_j refer to the a_j within the original sub-blocks, and the b_j' refer to the boundary bits $\mathbf{a_j}$ shown in boldface above. Note that because of the key addition, b_j and b_j' may differ.

2.5.5 DES S-Boxes

The so-called **S-boxes** (substitution boxes) are the core of the algorithm. For each of the eight sub-blocks, each consisting of six bits, there is a fixed **S-box**, namely a matrix with four rows and 16 columns. The rows are indexed by bit strings of length 2, the columns by bit strings of length 4 in ascending binary order. Each row of the matrix also contains all bit strings of length 4, but in rather jumbled order. For example, Table 2.1 shows the third S-box, matching the third 6-bit subblock $b_8' b_9 b_{10} b_{11} b_{12} b_{13}'$. To illustrate how S-boxes work, consider the specific example $b_8' b_9 b_{10} b_{11} b_{12} b_{13}' = 101011$. The two outer bits $b_8' b_{13}' = 11$ denote the row of the matrix, the four inner bits $b_9\, b_{10}\, b_{11}\, b_{12} = 0101$ decide the column. Therefore, the S-box returns the bit string 1001 and thus decides that the string $b_9 b_{10} b_{11} b_{12} = 0101$ is substituted by the string $c_9 c_{10} c_{11} c_{12} = 1001$. The border bits $b_8' b_{13}'$ are discarded again.

The output of the eight S-boxes finally results in a bit string $c_1\, c_2 \ldots c_{32}$ of length 32. For the sake of completeness, all eight S-boxes of DES are listed in Table 2.2.

2.5.6 DES Permutation

Finally, the output string $c_1\, c_2 \ldots . c_{32}$ of the eight S-boxes of bit length 32 is subjected to **permutation** π. It is permuted in the order according to Table 2.3, i.e. the bit from position

Table 2.1 Third S-box of DES with example

S3	0000	0001	0010	0011	0100	***0101***	0110	0111	1000	1001	1010	1011	1100	1101	1110	1111
00	1010	0000	1001	1110	0110	0011	1111	0101	0001	1101	1100	0111	1011	0100	0010	1000
01	1101	0111	0000	1001	0011	0100	0110	1010	0010	1000	0101	1110	1100	1011	1111	0001
10	1101	0110	0100	1001	1000	1111	0011	0000	1011	0001	0010	1100	0101	1010	1110	0111
11	0001	1010	1101	0000	0110	***1001***	1000	0111	0100	1111	1110	0011	1011	0101	0010	1100

Table 2.2 All eight S-boxes S1 to S8 of DES

S1	**0000**	**0001**	**0010**	**0011**	**0100**	**0101**	**0110**	**0111**	**1000**	**1001**	**1010**	**1011**	**1100**	**1101**	**1110**	**1111**
00	1110	0100	1101	0001	0010	1111	1011	1000	0011	1010	0110	1100	0101	1001	0000	0111
01	0000	1111	0111	0100	1110	0010	1101	0001	1010	0110	1100	1011	1001	0101	0011	1000
10	0100	0001	1110	1000	1101	0110	0010	1011	1111	1100	1001	0111	0011	1010	0101	0000
11	1111	1100	1000	0010	0100	1001	0001	0111	0101	1011	0011	1110	1010	0000	0110	1101

S2	**0000**	**0001**	**0010**	**0011**	**0100**	**0101**	**0110**	**0111**	**1000**	**1001**	**1010**	**1011**	**1100**	**1101**	**1110**	**1111**
00	1111	0001	1000	1110	0110	1011	0011	0100	1001	0111	0010	1101	1100	0000	0101	1010
01	0011	1101	0100	0111	1111	0010	1000	1110	1100	0000	0001	1010	0110	1001	1011	0101
10	0000	1110	0111	1011	1010	0100	1101	0001	0101	1000	1100	0110	1001	0011	0010	1111
11	1101	1000	1010	0001	0011	1111	0100	0010	1011	0110	0111	1100	0000	0101	1110	1001

S3	**0000**	**0001**	**0010**	**0011**	**0100**	**0101**	**0110**	**0111**	**1000**	**1001**	**1010**	**1011**	**1100**	**1101**	**1110**	**1111**
00	1010	0000	1001	1110	0110	0011	1111	0101	0001	1101	1100	0111	1011	0100	0010	1000
01	1101	0111	0000	1001	0011	0100	0110	1010	0010	1000	0101	1110	1100	1011	1111	0001
10	1101	0110	0100	1001	1000	1111	0011	0000	1011	0001	0010	1100	0101	1010	1110	0111
11	0001	1010	1101	0000	0110	1001	1000	0111	0100	1111	1110	0011	1011	0101	0010	1100

S4	**0000**	**0001**	**0010**	**0011**	**0100**	**0101**	**0110**	**0111**	**1000**	**1001**	**1010**	**1011**	**1100**	**1101**	**1110**	**1111**
00	0111	1101	1110	0011	0000	0110	1001	1010	0001	0010	1000	0101	1011	1100	0100	1111
01	1101	1000	1011	0101	0110	1111	0000	0011	0100	0111	0010	1100	0001	1010	1110	1001
10	1010	0110	1001	0000	1100	1011	0111	1101	1111	0001	0011	1110	0101	0010	1000	0100
11	0011	1111	0000	0110	1010	0001	1101	1000	1001	0100	0101	1011	1100	0111	0010	1110

S5	**0000**	**0001**	**0010**	**0011**	**0100**	**0101**	**0110**	**0111**	**1000**	**1001**	**1010**	**1011**	**1100**	**1101**	**1110**	**1111**
00	0010	1100	0100	0001	0111	1010	1011	0110	1000	0101	0011	1111	1101	0000	1110	1001
01	1110	1011	0010	1100	0100	0111	1101	0001	0101	0000	1111	1010	0011	1001	1000	0110
10	0100	0010	0001	1011	1010	1101	0111	1000	1111	1001	1100	0101	0110	0011	0000	1110
11	1011	1000	1100	0111	0001	1110	0010	1101	0110	1111	0000	1001	1010	0100	0101	0011

S6	**0000**	**0001**	**0010**	**0011**	**0100**	**0101**	**0110**	**0111**	**1000**	**1001**	**1010**	**1011**	**1100**	**1101**	**1110**	**1111**
00	1100	0001	1010	1111	1001	0010	0110	1000	0000	1101	0011	0100	1110	0111	0101	1011
01	1010	1111	0100	0010	0111	1100	1001	0101	0110	0001	1101	1110	0000	1011	0011	1000
10	1001	1110	1111	0101	0010	1000	1100	0011	0111	0000	0100	1010	0001	1101	1011	0110
11	0100	0011	0010	1100	1001	0101	1111	1010	1011	1110	0001	0111	0110	0000	1000	1101

S7	**0000**	**0001**	**0010**	**0011**	**0100**	**0101**	**0110**	**0111**	**1000**	**1001**	**1010**	**1011**	**1100**	**1101**	**1110**	**1111**
00	0100	1011	0010	1110	1111	0000	1000	1101	0011	1100	1001	0111	0101	1010	0110	0001
01	1101	0000	1011	0111	0100	1001	0001	1010	1110	0011	0101	1100	0010	1111	1000	0110
10	0001	0100	1011	1101	1100	0011	0111	1110	1010	1111	0110	1000	0000	0101	1001	0010
11	0110	1011	1101	1000	0001	0100	1010	0111	1001	0101	0000	1111	1110	0010	0011	1100

S8	**0000**	**0001**	**0010**	**0011**	**0100**	**0101**	**0110**	**0111**	**1000**	**1001**	**1010**	**1011**	**1100**	**1101**	**1110**	**1111**
00	1101	0010	1000	0100	0110	1111	1011	0001	1010	1001	0011	1110	0101	0000	1100	0111
01	0001	1111	1101	1000	1010	0011	0111	0100	1100	0101	0110	1011	0000	1110	1001	0010
10	0111	1011	0100	0001	1001	1100	1110	0010	0000	0110	1010	1101	1111	0011	0101	1000
11	0010	0001	1110	0111	0100	1010	1000	1101	1111	1100	1001	0000	0011	0101	0110	1011

Table 2.3 Permutation π within a DES round

16	7	20	21	29	12	28	17	1	15	23	26	5	18	31	10	2	8	24	14	32	27	3	9	19	13	30	6	22	11	4	25

16 is put in position 1, the bit from position 7 is put in position 2, and so on. This serves the equal distribution of the bits from round i to the S-boxes in round i + 1.

2.5.7 DES Key Selection

We now return to the selection and determination of the round key k_i. For this purpose, the 64-bit total key to be exchanged secretly is first written into the scheme of Table 2.4. The right column contains the parity check bits, which are now omitted.

The remaining bits are divided into two registers C (framed in bold on the left) and D, in the order determined according to Table 2.5. This is called PC-1 (Permuted Choice 1).

In each round, 24 bits are selected from each of the two registers, always the same ones and always in the same order. Table 2.6 shows the selection and the sequence. Here, the numbering refers to the position numbers shown in italics in Table 2.5. This is called PC-2 (Permuted Choice 2).

However, in order to obtain different round keys k_i for each round, the values in registers C and D are cyclically shifted to the left after each round according to Table 2.5, namely by one position after rounds 1, 2, 9 and 16 and by two positions after the remaining rounds. Therefore, during the 16 rounds, a total of 28 shift operations are performed, so that the registers are in their initial state again afterwards. Therefore, the next 64-bit block can be enciphered without reloading the key.

2.5.8 Security of the DES

The many operations and permutations per round do not serve the actual encryption, because they are all publicly known and realized in DES programs. Rather with them the secret key is to be mixed as powerfully as possible into the plaintext. For example, it can be shown that after only five rounds of DES, every bit depends on every plaintext bit and every key bit, i.e., DES produces a high degree of diffusion. This is also one reason why, with the exception of the brute-force attack, there is no other “real practical” attack on DES to date, not even differential and linear cryptanalysis, which emerged in the early 1990s and are generally applicable to iterative block ciphers and, in particular, Feistel

Table 2.4 DES key scheme with parity check bits (right column)

1	2	3	4	5	6	7	*8*
9	10	11	12	13	14	15	*16*
17	18	19	20	21	22	23	*24*
25	26	27	28	29	30	31	*32*
33	34	35	36	37	38	39	*40*
41	42	43	44	45	46	47	*48*
49	50	51	52	53	54	55	*56*
57	58	59	60	61	62	63	*64*

Table 2.5 PC-1 with registers C and D

	1	2	3	4	5	6	7	8	9	10	11	12	13	14	15	16	17	18	19	20	21	22	23	24	25	26	27	28
C	57	49	41	33	25	17	9	1	58	50	42	34	26	18	10	2	59	51	43	35	27	19	11	3	60	52	44	36
D	63	55	47	39	31	23	15	7	62	54	46	38	30	22	14	6	61	53	45	37	29	21	13	5	28	20	12	4
	29	30	31	32	33	34	35	36	37	38	39	40	41	42	43	44	45	46	47	48	49	50	51	52	53	54	55	56

Table 2.6 PC-2 with the key selection and sequence

14	17	11	24	1	5	3	28	15	6	21	10	23	19	12	4	26	8	16	7	27	20	13	2
41	52	31	37	47	55	30	40	51	45	33	48	44	49	39	56	34	53	46	42	50	36	29	32

ciphers. We will now explain these two important attack strategies in more detail using DES. This will get a bit tricky, but is not absolutely necessary for further understanding. Therefore, both topics can also just be "skimmed over" or even skipped.

2.5.9 Differential Cryptanalysis Using the Example of DES

Differential cryptanalysis is a special chosen plaintext attack (Sect. 2.1). The attacker encrypts two plaintext blocks m and m′ with a self-selected difference (= sum) $m \oplus m'$ and learns at least the difference $c \oplus c'$ of the ciphertext blocks c and c′. When executed multiple times, he or she can thus examine the effects of differences in plaintext blocks on differences in the associated ciphertext blocks. This allows the probability of keys, and hence the most likely key, to be determined.

So we want to explain the procedure at the example of DES, but for the sake of clearness only at one round, and only at that part, which operates on the right block with a round key k. Let two 32-bit strings R and R′ as well as the difference $F(R,k) \oplus F(R',k)$ be known. On the way through the DES round, the only unknown is the key k, which has to be determined or narrowed down. Let B and B′ be the input strings to the S-boxes belonging to R and R′, respectively, and C and C′ be the corresponding output strings. Then $B = \varepsilon(R) \oplus k$ and $B' = \varepsilon(R') \oplus k$ with expansion mapping ε, and consequently $B \oplus B' = (\varepsilon(R) \oplus k) \oplus (\varepsilon(R') \oplus k) = \varepsilon(R) \oplus \varepsilon(R') = \varepsilon(R \oplus R')$. In particular, therefore, although B and B′ are not known individually, at least their difference is known. Furthermore, $C \oplus C' = \pi^{-1}(F(R,k)) \oplus \pi^{-1}(F(R',k)) = \pi^{-1}(F(R,k) \oplus F(R',k))$, such that with the help of the inverse permutation π^{-1} of π, the difference $C \oplus C'$ is also known. If it is now possible to restrict B or B′, then k is also determined accordingly because of $k = B \oplus \varepsilon(R) = B' \oplus \varepsilon(R')$.

We use the abbreviation $E = \varepsilon(R)$ und $E' = \varepsilon(R')$ and explain the procedure on the basis of the first S-box S1. For this the index 1 may denote the part of the respective bitstrings, which refers to S1. Thus, with this designation, $B_1 \oplus B_1'$ and $C_1 \oplus C_1'$ are also known in particular, and $k_1 = B_1 \oplus E_1$ holds. As a concrete example [Hau1] now let $R = 00101 * \ldots * 1$, $R' = 10001 * \ldots * 1$ and $C_1 \oplus C_1' = 0110$. Then $E_1 = 100101$, $E_1' =$

110001 and consequently $B_1 \oplus B_1' = E_1 \oplus E_1' = 010100$. Now one determines all 6-bit pairs X and X′ with $X \oplus X' = 010100$ in such a way that these result in 4-bit pairs Y and Y′ with difference $Y \oplus Y' = 0110$ when passing through the S-box S1. After a little calculation one obtains for B_1 the four possibilities 100010, 110110, 101010, 111110 and consequently for $k_1 = B_1 \oplus E_1$ the four possibilities 000111, 010011, 001111, 011011. If one repeats the procedure for other R and R′, one can thus further narrow down k_1. Accordingly, one proceeds to determine the total round key $k = k_1 \ldots k_8$ with the other S-boxes.

To shorten the calculations, one can of course keep corresponding tables for all eight S-boxes and all possible differences. In Table 2.7 for our example of the S-Box S1 with input difference $X \oplus X' = 010100$ to all output differences $Y \oplus Y'$ the number of the different possibilities is listed.

With more than one DES round, however, differential cryptanalysis becomes more and more complex; with 16 rounds it is not significantly more effective than a brute-force attack. Although not officially published until 1991 by **Eli Biham** (born 1960) and **Adi Shamir** (born 1952), the DES developers nevertheless already knew the underlying method.

2.5.10 Linear Cryptanalysis Using the Example of DES

We have subliminally used the equations $\varepsilon(R \oplus R') = \varepsilon(R) \oplus \varepsilon(R')$ and $\pi(C \oplus C') = \pi(C) \oplus \pi(C')$ for the expansion ε and the permutation π in differential

Table 2.7 Number of output differences of S-Box S1 with input difference 010100

Output difference	Number
0000	0
0001	8
0010	8
0011	0
0100	10
0101	0
0110	4
0111	2
1000	8
1001	2
1010	2
1011	4
1100	4
1101	8
1110	4
1111	0

cryptanalysis. Indeed, both ε and π are so-called **linear transformations**. However, the S-boxes are highly nonlinear transformations, as can be seen, for example, from Table 2.7. If the S-box S1 were linear, all 64 input pairs with input difference 010100 would lead to the same output difference. Therefore, the entire DES algorithm is also nonlinear. **Linear cryptanalysis** is a known-plaintext attack (Sect. 2.1) that attempts to "linearly approximate" a block cipher as optimally as possible in order to determine the key from a sufficient number of plaintext/ciphertext pairs, at least partially and with a certain probability. The method was developed by **Mitsuru Matsui** (born 1961) in 1993.

We want to explain the basic idea again by the example of the DES, and again only by one round and quite concretely by the S-box S5. Let $B_5 = b_1 \ldots b_6$ be an input string and $C_5 = c_1 \ldots c_4$ the corresponding output string. For the **linear approximation** of the S-box S5 one searches for bit strings $u_1 \ldots u_6$ and $v_1 \ldots v_4$ in such a way that one of the two following equations is fulfilled for far more than half of all $2^6 = 64$ possible input strings B_5 with the corresponding output strings C_5 and thus has a high probability:

$$u_1 \bullet b_1 + \ldots + u_6 \bullet b_6 = v_1 \bullet c_1 + \ldots + v_4 \bullet c_4$$
$$u_1 \bullet b_1 + \ldots + u_6 \bullet b_6 = v_1 \bullet c_1 + \ldots + v_4 \bullet c_4 + 1$$

To do this, one sets up a table that contains, for all values of $u_1 \ldots u_6$ and $v_1 \ldots v_4$, the indication of how often the first of the two equations is satisfied for input string B_5 with output string C_5.

For example, for the values $u_1 \ldots u_6 = 010000$ and $v_1 \ldots v_4 = 1111$, when passing through the S-box S5 one finds [Fra] that the first equation $u_1 \bullet b_1 + \ldots + u_6 \bullet b_6 = b_2 = c_1 + \ldots + c_4 = v_1 \bullet c_1 + \ldots + v_4 \bullet c_4$ holds only in 12 cases out of a total of 64. So, conversely, the second equation $b_2 = c_1 + \ldots + c_4 + 1$ holds in 52 out of 64 cases and this therefore has a probability of 0.81. Considering now again the expansion $\varepsilon(R_5) = E_5 = e_1 \ldots e_6$ and the key portion $K_5 = k_1 \ldots k_6$ to the S-box S5, in a known-plaintext attack with n plaintext/chiffretext pairs, the respective bit strings E_5 and C_5 are known, and in this case $B_5 = E_5 \oplus K_5$ holds with previously unknown K_5. Therefore, our linear approximation yields $e_2 + k_2 = c_1 + \ldots + c_4 + 1$, and k_2 is with high probability the bit 0 or 1 for which the equation is correct for more than n/2 of the plaintext/chiffretext pairs E_5 and C_5.

For $u_1 \ldots u_6 = 111111$ and $v_1 \ldots v_4 = 0100$, one can verify that $u_1 \bullet b_1 + \ldots + u_6 \bullet b_6 = b_1 + \ldots + b_6 = c_2 = v_1 \bullet c_1 + \ldots + v_4 \bullet c_4$ holds in 46 out of a total of 64 cases, i.e. with probability 0.72. This yields $e_1 + \ldots + e_6 + k_1 + \ldots + k_6 = c_2$, and one in turn determines from this $k_1 + \ldots + k_6$ as the bit for which the equation is correct for more than n/2 of the plaintext/ciphertext pairs. The two results for the key bits $K_5 = k_1 \ldots k_6$ can now be combined with each other or further linear approximations of high probability can be determined and used.

With more than one DES round, the creation of linear approximations becomes more and more difficult and thus the linear cryptanalysis more and more complex. The method was not known to the developers of DES in contrast to differential cryptanalysis. Therefore also the S-Boxes are not completely optimized concerning linear cryptanalysis.

2.5.11 Brute-Force Attack and Triple-DES

By design, DES would actually only have required an effective key length of 48 bits. However, even when it was first standardized, this tended to be insecure because of the possibility of brute-force attacks. Nevertheless, the relatively short key of effectively 56 bits proved to be DES's greatest weakness. While a complete key search was hardly conceivable at the time of DES's introduction, it came within immediate reach in the 1990s. With large networked computers, it was possible to get into the day and even hour range. To increase the effective key length, an obvious approach is to encrypt multiple times with DES. However, one may consider that even double DES encryption hardly provides more security. To get around this problem, the so-called **Triple-DES** has been introduced. Here, one uses three DES algorithms with two independent keys k_1 and k_2, first and finally applying the DES cipher $E_{DES}(\bullet, k_1)$ with the key k_1, but in between applying the DES decipher $D_{DES}(\bullet, k_2)$ with the key k_2. Triple-DES then has an effective key length of 112 bits.

$$\text{Plaintext m} \rightarrow \text{Ciphertext c} = E_{DES}\left(D_{DES}\left(E_{DES}\left(m,k_1\right),k_2\right),k_1\right)$$

Triple DES was and is implemented in many practical applications for the encryption of data requiring protection. However, it has been successively replaced by the more modern AES (Sect. 2.8), or at least AES is offered as an alternative.

2.6 Operating Modes of Block Ciphers

2.6.1 Electronic Codebook Mode

Let $E = E(\bullet, k)$ be an arbitrary block cipher with key k and with binary input and output blocks, e.g. DES or Triple-DES. Furthermore, let $D = D(\bullet, k')$ be the corresponding decryption scheme with possibly different decryption key k'. Let us again denote the blocks of plaintext by $m_1, m_2, \ldots, m_n$, with the last block m_n padded to the same length as the others, if necessary. Then the blocks c_i of the ciphertext are computed according to $c_i = E(m_i, k)$. This use of a block cipher, namely according to its very own definition, as it

Fig. 2.8 ECB mode of a block cipher E

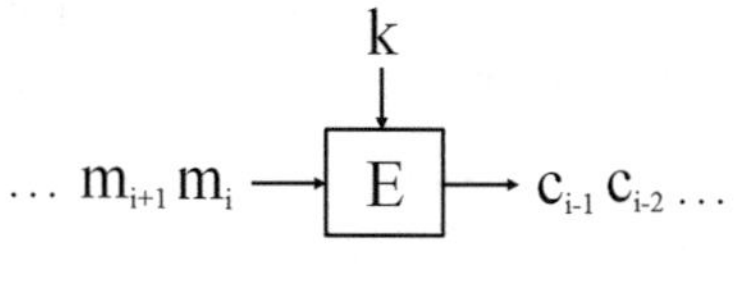

$$\ldots m_{i+1}\, m_i \longrightarrow \boxed{E} \longrightarrow c_{i-1}\, c_{i-2} \ldots \quad (\text{key } k \text{ into } E)$$

is also visualized in Fig. 2.8, is called **electronic codebook mode (ECB)**. This means that identical blocks are always encrypted identically. This preserves large-scale plaintext patterns, and the frequency of identical plaintext areas is only inadequately disguised. Thus, the ECB mode provides ideal attack conditions for statistical analyses, as we have used several times for historical ciphers. Another disadvantage of the ECB mode is that the receiver of the ciphertext cannot necessarily detect whether an attacker has deleted, swapped, or even added blocks during data transmission. In general, the ECB mode should therefore only be used for short messages with few blocks. Decryption in ECB mode is performed according to $m_i = D(c_i, k')$.

2.6.2 Cipher Block Chaining Mode

Plaintext patterns can be destroyed using contextual encryption. In **cipher block chaining mode (CBC)**, one adds the previous ciphertext block to the current plaintext block and then encrypts the result. Thus, the blocks c_i of the ciphertext are calculated according to $c_i = E(m_i \oplus c_{i-1}, k)$. However, since one does not yet have a ciphertext block available for the first plaintext block m_1, one uses an initialization block c_0, which is sent to the receiver together with the entire ciphertext. Figure 2.9 again visualizes the procedure. Decryption is performed according to $m_i = D(c_i, k') \oplus c_{i-1}$.

2.6.3 Cipher Feedback Mode

In **cipher feedback mode (CFB)**, one computes the blocks c_i of the ciphertext according to $c_i = m_i \oplus E(c_{i-1}, k)$. Thus, it is a stream cipher where the block cipher E is used to generate a context-dependent pseudo-random sequence that is added to the plaintext. Again, an initialization block c_0 is required, which is sent to the receiver along with the entire ciphertext. Figure 2.10 again visualizes the procedure.

Because of $c_i \oplus E(c_{i-1}, k) = m_i \oplus E(c_{i-1}, k) \oplus E(c_{i-1}, k) = m_i$ one does not even need D for deciphering, but computes the plaintext according to $m_i = c_i \oplus E(c_{i-1}, k)$. Here, as with all stream ciphers, the receiver has the additional advantage that he or she does not have to wait for the entire ciphertext block c_i, but can decrypt it bit by bit.

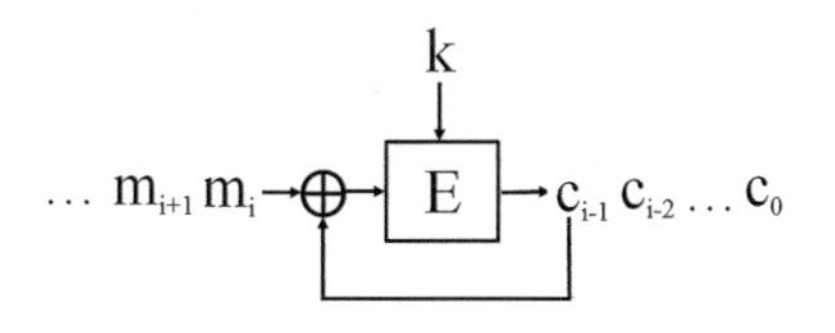

Fig. 2.9 CBC mode of a block cipher E

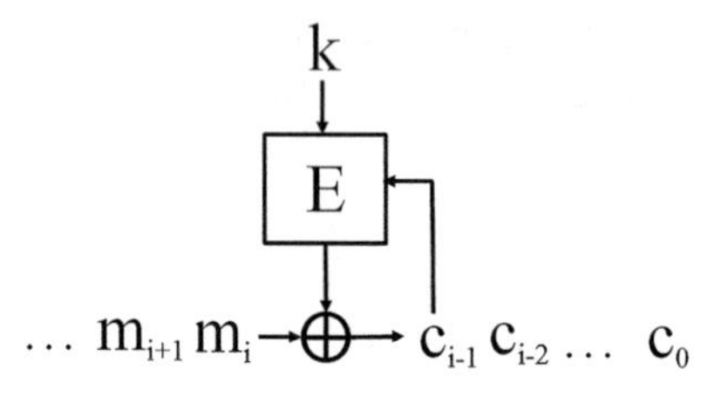

Fig. 2.10 CFB mode of a block cipher E

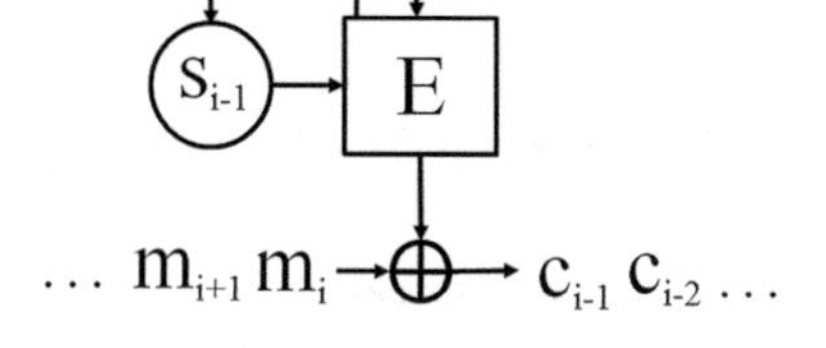

Fig. 2.11 OFB mode of a block cipher E

2.6.4 Output Feedback Mode

However, it is also possible to design a stream cipher based on the block cipher E in a context-independent manner, which has the advantage that the pseudo-random sequence can be calculated in advance. For this purpose, the sender and receiver agree on an initial value s_0 of the same length as m_i. In **output feedback mode (OFB)**, the ciphertext block c_i is then determined according to $c_i = m_i \oplus E(s_{i-1}, k)$, so here the pseudo-random sequence $s_i = E(s_{i-1}, k)$ has no reference to the context. Figure 2.11 shows the procedure. Decryption is again bitwise according to $m_i = c_i \oplus E(s_{i-1}, k)$.

2.6.5 Counter Mode

Finally, we want to describe the **counter mode (CTR)**, where the encryption of the plaintext block m_i depends on its position i in the text m. For this purpose, one writes the position $i = i_0 \bullet 2^0 + i_1 \bullet 2^1 + i_2 \bullet 2^2 + i_3 \bullet 2^3 + \ldots + i_{b-1} \bullet 2^{b-1}$ as a binary expansion with bits i_j and the block length b of m_i and again agrees on a base value s_0 of length b. Now identifying i with the bit string $i_0\, i_1\, i_2\, i_3 \ldots i_{b-1}$, one can add s_0 and i bitwise $\oplus$ and derive a context-dependent stream cipher where the ciphertext block c_i is computed according to $c_i = m_i \oplus E(s_0 \oplus i, k)$. However, despite context dependency, one can calculate the pseudo-random sequence in advance here as in the OFB mode. Figure 2.12 illustrates the CTR mode. If one has reached position $i = 2^b$, one simply counts from the beginning again. Decoding in the CTR mode is done bit by bit according to $m_i = c_i \oplus E(s_0 \oplus i, k)$.

The operating modes were first standardized for use with DES in 1981, but are of course used with other block ciphers as well.

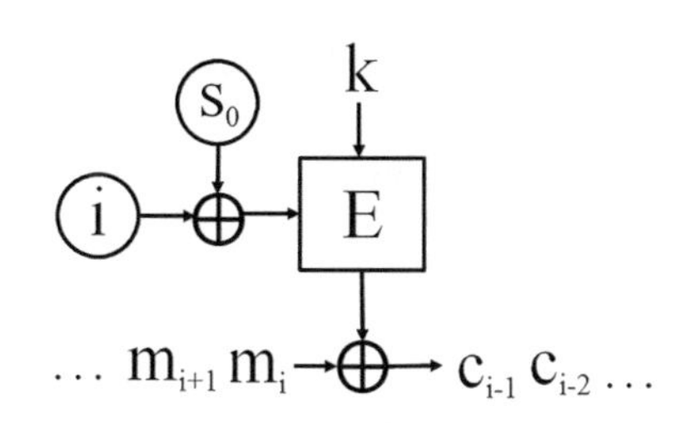

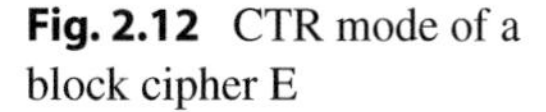
Fig. 2.12 CTR mode of a block cipher E

2.7 UMTS/LTE Mobile Communications and Digital Television

2.7.1 The UMTS/LTE Mobile Communications Standard

We will now take up the encryption procedure for the GSM mobile communications standard (Sect. 2.3). In the course of the 1990s, **UMTS** (Universal Mobile Telecommunications System) was developed as a 3. generation **(3G)** mobile communications standard with significantly higher data transmission rates than GSM. UMTS includes additional services such as e-mail and Internet. UMTS has been commercially available in Germany since 2004 and there are now UMTS networks in over 100 countries. In the meantime, **LTE** (Long Term Evolution) has already been launched as a 4.generation **(4G)** mobile communications standard, but it has a similar architecture to UMTS. In 2010, the first LTE licenses were auctioned in Germany and the first LTE transmission masts were put into operation.

2.7.2 A5 Cipher of Versions A5/3 and A5/4

While the old version A5/1 is still widely used in many GSM networks and is only gradually being replaced by A5/3, the A5/4 version is already implemented for data encryption in UMTS and LTE. Both are fundamentally different from A5/1. It is the Japanese **KASUMI cipher** (English: fog, mist), a variant of **MISTY1** from 1995. KASUMI is a Feistel cipher with 8 rounds on 64-bit blocks and a 128-bit key. This generates a pseudorandom sequence in a combination of CTR and OFB modes and is therefore operated as a stream cipher. Since we have already dealt in detail with DES and thus with by far the most important Feistel cipher, we will not give an explicit description of the round function [3GPP] for the KASUMI cipher.

The standardization of A5/3 has in fact a key with an effective key length of 64 bits. This is simply doubled to a key length of 128 bits for the KASUMI algorithm. One reason for this is that the key generation for GSM can be used unchanged for A5/3 and thus GSM can be upgraded to A5/3 more easily (Sect. 2.3). However, this means that A5/3 is just as vulnerable to brute-force attacks as DES. For this reason, **ETSI** (European Telecommunications Standards Institute) has also launched version A5/4, also with KASUMI cipher, but with an effective key length of 128 bits [ETSI2, WPA5A].

2.7.3 DVB and MPEG2

DVB (Digital Video Broadcasting) is a standard for the digital transmission of television programmes. There are different sub-standards for different transmission paths, which differ, among other things, in the modulation method: DVB-S for transmission via satellite, DVB-C for transmission via cable networks, DVB-T for transmission via terrestrial transmitters. DVB-S and DVB-C were ratified in 1994, DVB-T 3 years later. In the meantime, however, there is already a successor standard DVB2. The video and audio contents of DVB are transmitted by means of so-called **MPEG2 transport packets**. These are named after the **MPEG** (Moving Picture Experts Group), which has been creating various standards for video and audio formats since the late 1980s. Each MPEG2 transport packet consists of header data with controlling information as well as the actual payload data. For example, there is a 2-bit field that encodes a possible encryption, where 00 stands for unencrypted. The transport packets are reassembled during playback to form the so-called elementary stream, which ultimately generates the video and audio playback.

2.7.4 CSA Encryption

We will now look at DVB encryption for pay-TV channels. The procedure, which originated in 1994, is called **CSA** (Common Scrambling Algorithm). Each receiver requires a **CA module** (Condition Access) and an individual smart card (ICC). In addition to the MPEG2 transport packets, the DVB provider sends separate **ECM packets** (Entitlement Control Message) with which the keys for the decryption of the pay-TV channel are

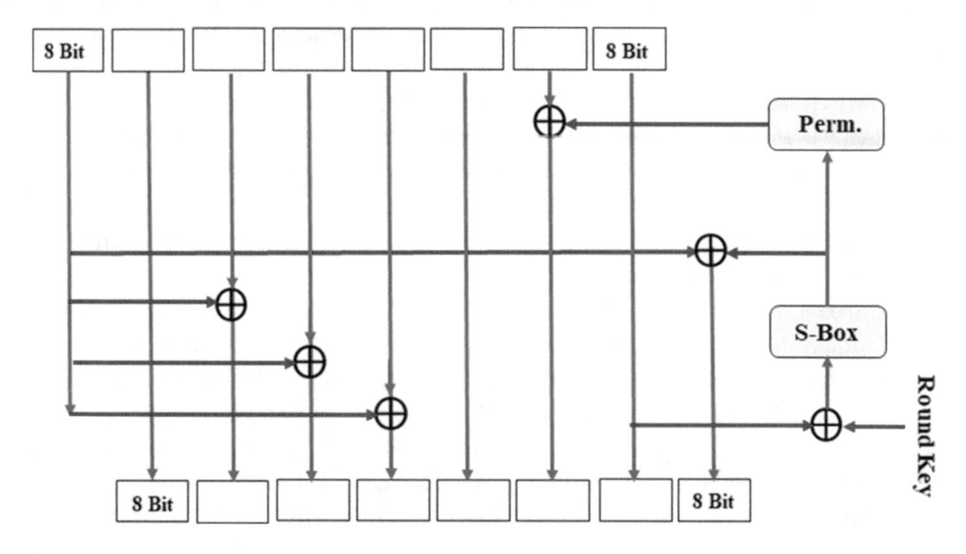

Fig. 2.13 Round function of the CSA block cipher (schematic)

transmitted. The CA module filters the ECM packets out of the data stream and uses the smart card to calculate the 64-bit key that is valid at that time.

The CSA encryption method itself consists of a combination of block cipher and stream cipher, with the block cipher being used for encryption first. This is not a Feistel cipher, but more generally an iterative substitution permutation cipher of 56 rounds on blocks of 64 bits, operated in CBC mode. The round function is shown schematically in Fig. 2.13, where the permutation, the substitution box, and the derivative of the round key are specified separately.

Following the block cipher, an additional complex stream cipher is used, which outputs two pseudo-random bits at each of its clock pulses, which are added to the bit stream to be encrypted [WPCAS]. CSA was kept secret for many years, contrary to the Kerckhoffs principle, but then became public knowledge in 2002. Although a brute-force attack initially appears feasible due to the small key length, it is hampered by the frequent change of the key in the ECM packets.

In 2013, ETSI standardized a successor procedure **CSA3**, which is based on the modern standard procedure AES (Sect. 2.8) and on an XRC cipher, which is again kept secret [ETSI1]. The AES cipher is operated with a key length of 128 bits in CBC mode. However, CSA3 is hardly used, and CSA therefore remains the dominant method for protecting pay-TV channels in DVB.

2.8 Advanced Encryption Standard AES

2.8.1 Rijndael Procedure and AES

In view of the growing threat of brute-force attacks on DES, the US standardization authority NIST launched a public tender for a successor procedure to DES in 1997. In contrast to the call for bids of the 1970s, as many as 15 applicants worldwide submitted bids in 1998, of which five procedures finally made it to the final round in 1999. These were the Feistel ciphers **MARS**, **RC6** and **Twofish** as well as **Serpent** and **Rijndael**, which were designed as iterative **substitution permutation ciphers**. In 2000, **Rijndael** was chosen by **Joan Daemen** (b. 1965) and **Vincent Rijmen** (b. 1970), and 2 years later their method was officially declared the Federal Information Processing Standard in the United States as **AES (Advanced Encryption Standard)**. Although all five methods were classified as secure by NIST, the speed advantage may ultimately have tipped the scales in favor of Rijndael.

The AES is therefore an iterative block cipher, but not a Feistel cipher. Since AES is currently by far the most important symmetric encryption method, the algorithm is to be described here to the extent that it conveys as concrete an understanding as possible. The AES in its standardized form has a block length of 128 bits and allows key lengths of 128, 192 or 256 bits. It consists of

- a preliminary round,
- 9, 11 or 13 normal rounds (each for key length 128, 192 or 256 bits) and
- a final round.

In the normal rounds, the following four modules are executed:

- SubByte
- ShiftRow
- MixColumn
- AddRoundKey

The preliminary round uses only AddRoundKey, and the final round does without MixColumn. Before we describe these four building blocks in a little more detail, we want to point out what is actually new about the AES cipher.

2.8.2 Addition and Multiplication of Bytes

Here again are the addition and multiplication tables for bits (Sect. 1.2):

+	0	1
0	0	1
1	1	0

·	0	1
0	0	0
1	0	1

As we already know, information units such as letters and pixels are usually interpreted as blocks of several bits, especially often as bytes with 8 bits. The question therefore arises whether it is also possible to add and multiply blocks of bits in a meaningful way. But what should "meaningful" mean in this context? It should mean that certain rules of calculation apply, which are absolutely necessary for further considerations. It is important, that you can not only add and multiply, but also subtract and divide. For this you need the following formal properties:

- Each bit block can be made 0…0 by adding a second bit block (so-called **additive inverse**, which corresponds to a subtraction). **0…0** is also called the 0-element.
- Each bit block not equal to 0…0 can be made 0…01 by multiplication with a second bit block (so-called **multiplicative inverse**, which corresponds to a division). 0…01 is also called the **1-element**.

In mathematics, such structures are also called a **field**. Let us look at the simple example of bit pairs. A first approach is to add and multiply the bits position by position. For our well-known bitwise addition $\oplus$ this works well, because if you add the same bit pair to a bit pair, the result is always 00. But for the multiplication, unfortunately, one suffers a shipwreck, because no matter what one multiplies 10 position by position, it never results in 01. So one has to define the multiplication more elaborate. Here are the desired useful addition and multiplication tables for bit pairs:

⊕	00	01	10	11
00	00	01	10	11
01	01	00	11	10
10	10	11	00	01
11	11	10	01	00

⊗	00	01	10	11
00	00	00	00	00
01	00	01	10	11
10	00	10	11	01
11	00	11	01	10

In a way it is an extension of the bit addition and multiplication, because this is found exactly in the left upper quarter of the table, related to the last position of the bit pair. Now there is also a second bit pair to 10, namely 11, for which the product 10 ⊗ 11 results in the 1-element 01. A corresponding multiplication ⊗ works also for bit blocks of any length, so especially also for bytes with their 8 bits. However, we refrain here from reproducing the byte multiplication table ⊗ with its 64 rows and columns.

Instead we want to explain briefly how to define this reasonable multiplication ⊗ conceptually for bit blocks of arbitrary length n. For this we number the positions of the bits on the right starting from 0 to n – 1 and set for abbreviation t = 0…0010. For i from 0 to n – 1 we now define $t^i = t \otimes t \otimes \ldots_i \ldots \otimes t$ as the bit string which has bit 1 exactly at position i and 0 otherwise. In particular, $t^0 = 0\ldots001$ is the 1-element, and $t^i \otimes t^j = (t \otimes t \ldots_i \ldots \otimes t) \otimes (t \otimes t \ldots_j \ldots \otimes t) = t^{i+j}$. Any bit strings, i.e. sums ⊕ of some t^i, are multiplied by the distributive rule. For example, for n = 8 we have $00000101 \otimes 00001110 = (t^2 \oplus t^0) \otimes (t^3 \oplus t^2 \oplus t^1) = t^5 \oplus t^4 \oplus t^3 \oplus t^3 \oplus t^2 \oplus t^1 = t^5 \oplus t^4 \oplus t^2 \oplus t^1 = 00110110$. But wait: This multiplication rule ⊗ makes sense only if the exponents of t are at most n – 1. So you need some kind of recursion formula for t^n. But this formula is not so easy to construct for arbitrary n. For bit pairs, for example, the formula, would be $t^2 = t^1 \oplus t^0 = 11$, and for bit triples one can use $t^3 = t^1 \oplus t^0 = 011$. The reader is asked to check this for bit-pairs in the multiplication table above and to create the multiplication table for bit-triples with its eight rows and columns. For bytes at any rate one can use as one of several possibilities the recursion formula $t^8 = t^4 \oplus t^3 \oplus t^1 \oplus t^0 = 00011011$, and it is exactly this formula which is used for AES. Mathematically speaking, bytes thus form a field [Man, Wil, Buc].

2.8.3 AES State Matrix

What hence is essentially new compared to DES is that AES is not designed for bit structures, but for byte structures and their addition and multiplication. A plaintext block of

AES has 128 bits, i.e. 128/8 = 16 bytes. Each such block $a_1 \ldots a_{16}$ of 16 bytes is read for encryption column by column into a matrix with four rows and four columns:

$$\begin{pmatrix} a_1 & a_5 & a_9 & a_{13} \\ a_2 & a_6 & a_{10} & a_{14} \\ a_3 & a_7 & a_{11} & a_{15} \\ a_4 & a_8 & a_{12} & a_{16} \end{pmatrix}$$

On the basis of this so-called **state matrix**, all mapping modules are now defined.

2.8.4 AES Subbyte

The block **SubByte** is the actual substitution part in the AES procedure. From our preliminary consideration we already know that every byte a_i, which is not equal to 00000000, has a multiplicative inverse a_i^{-1}, for which $a_i \otimes a_i^{-1} = 00000001$ therefore applies. So for each byte a_i in the state matrix, we first compute $b_i = a_i^{-1}$, if a_i is not equal to 00000000, and set $b_i = 00000000$ for $a_i = 00000000$. Then we write each byte $b_i = \beta_7^{(i)} \ldots \beta_0^{(i)}$ again as a bit string of length 8 and transform the bits $\beta_7^{(i)},\ldots,\beta_0^{(i)}$ according to

$$\begin{matrix} \beta_0^{(i)} & \rightarrow & \beta_{0'}^{(i)} = \lambda_{00} \cdot \beta_0^{(i)} + & \ldots & +\lambda_{07} \cdot \beta_7^{(i)} + \delta_0 \\ \vdots & & \vdots & & \vdots \\ \beta_7^{(i)} & \rightarrow & \beta_{7'}^{(i)} = \lambda_{70} \cdot \beta_0^{(i)} + & \ldots & +\lambda_{77} \cdot \beta_7^{(i)} + \delta_7 \end{matrix}$$

Here, the bits λ_{jl} and δ_j are determined by the AES algorithm as follows:

$$\left(\lambda_{jl}\right) = \begin{pmatrix} 1 & 0 & 0 & 0 & 1 & 1 & 1 & 1 \\ 1 & 1 & 0 & 0 & 0 & 1 & 1 & 1 \\ 1 & 1 & 1 & 0 & 0 & 0 & 1 & 1 \\ 1 & 1 & 1 & 1 & 0 & 0 & 0 & 1 \\ 1 & 1 & 1 & 1 & 1 & 0 & 0 & 0 \\ 0 & 1 & 1 & 1 & 1 & 1 & 0 & 0 \\ 0 & 0 & 1 & 1 & 1 & 1 & 1 & 0 \\ 0 & 0 & 0 & 1 & 1 & 1 & 1 & 1 \end{pmatrix} \text{ and } \left(\delta_j\right) = \begin{pmatrix} 1 \\ 1 \\ 0 \\ 0 \\ 0 \\ 1 \\ 1 \\ 0 \end{pmatrix}$$

The module SubByte then replaces each byte a_i in the state matrix with the byte $a_i^{'} = \beta_{7'}^{(i)} \ldots \beta_{0'}^{(i)}$ determined in this way.

2.8.5 AES ShiftRow

The module **ShiftRow** changes the rows of the state matrix. The first row remains unchanged, the second row is cyclically shifted to the left by one place, the third row by two places and the fourth row by three places. In this way, each byte a_i' of the state matrix is converted into a byte a_i''.

2.8.6 AES MixColumn

The module **MixColumn** changes the columns of the state matrix. For abbreviation we write e = 00000001 (the 1-element), t = 00000010 and s = e ⊕ t = 00000011. Then the elements a_1''', a_2''', a_3''' und a_4''' of the new first column of the state matrix are calculated as

$$\begin{aligned} a_1''' &= t \otimes a_1'' \oplus s \otimes a_2'' \oplus e \otimes a_3'' \oplus e \otimes a_4'' \\ a_2''' &= e \otimes a_1'' \oplus t \otimes a_2'' \oplus s \otimes a_3'' \oplus e \otimes a_4'' \\ a_3''' &= e \otimes a_1'' \oplus e \otimes a_2'' \oplus t \otimes a_3'' \oplus s \otimes a_4'' \\ a_4''' &= s \otimes a_1'' \oplus e \otimes a_2'' \oplus e \otimes a_3'' \oplus t \otimes a_4'' \end{aligned}$$

The new elements a_5''', a_6''', a_7''' and a_8''' of the second column, a_9''', a_{10}''', a_{11}''' and a_{12}''' of the third column and a_{13}''', a_{14}''', a_{15}''' and a_{16}''' of the fourth column of the state matrix are calculated in the same way.

2.8.7 AES AddRoundKey

Now, of course, the secret key k must also come into play. Since you want to use different keys for each round, you construct them successively on the basis of the 128-, 192-, or 256-digit AES key. We will explain the procedure using the example of a 128-digit key. To do this, the key k is first divided into four blocks k_0, k_1, k_2 and k_3 of 32 bits each. The block **AddRoundKey** then adds the actual AES key $k = k_0\, k_1\, k_2\, k_3$ as a round key bit by bit ⊕ to the plaintext block $a_1\, a_2 \ldots a_{15}\, a_{16}$ in the preliminary round.

For the j. round one recursively derives the following four 32-bit blocks from k:

$$\begin{aligned} k_{4j} &= k_{4j-4} \oplus T\left(k_{4j-1}\right) \\ k_{4j+1} &= k_{4j-3} \oplus k_{4j} \\ k_{4j+2} &= k_{4j-2} \oplus k_{4j+1} \\ k_{4j+3} &= k_{4j-1} \oplus k_{4j+2} \end{aligned}$$

Here the transformation T of the 32-bit block k_{4j-1} must still be described. However, this again consists of four bytes, say $k_{4j-1} = c_1^{(j)} c_2^{(j)} c_3^{(j)} c_4^{(j)}$ with the bytes $c_1^{(j)}$ to $c_4^{(j)}$. Then T transforms these bytes according to

$$\begin{aligned} c_1^{(j)} &\rightarrow S\left(c_2^{(j)}\right) \oplus t^{j-1} \\ c_2^{(j)} &\rightarrow S\left(c_3^{(j)}\right) \\ c_3^{(j)} &\rightarrow S\left(c_4^{(j)}\right) \\ c_4^{(j)} &\rightarrow S\left(c_1^{(j)}\right) \end{aligned}$$

with the transformation S described at the module SubByte and with the byte $t = 00000010$.

In the j. round, the 128-bit round key $k_{4j} \| k_{4j+1} \| k_{4j+2} \| k_{4j+3}$ formed by stringing together k_{4j}, k_{4j+1}, k_{4j+2} and k_{4j+3} is added bit by bit $\oplus$ to the concatenation $a_1''' \| a_2''' \| \ldots \| a_{15}''' \| a_{16}'''$ of the entries a_1''' to a_{16}''' of the state matrix in the AddRoundKey module.

2.8.8 AES Decryption

AES is not a Feistel cipher, which can also be used for deciphering in the same way only with reversed order of the round keys. But it is not hard to see, that all AES modules are invertible, i. e. can be inverted again. Again, one needs the same round keys, only in reverse order.

2.8.9 Security of the AES

Of course, it is not surprising that numerous cryptanalytic attacks have been carried out on AES. However, the method is secure against all attacks known to date, e.g. also against differential and linear cryptanalysis. The inversion of the bytes in the module SubBytes makes the method highly complex, the modules ShiftRow and MixColumn cause a high confusion and diffusion. However, it is debatable whether the simple algebraic design could be a weakness of AES and thus a possible point of attack.

2.9 Hard Disk and ZIP Archive

2.9.1 Hard Disk

The magnetic storage medium **hard disk** has been the most important mass storage medium for many decades. Hard disk drives are installed in computers, but are also offered as external drives. The write and simultaneously read head of the write finger is basically a small electromagnet. It magnetizes tiny areas of the disk surface differently and thus writes the data to the hard disk. Conversely, when reading, the changes in the magnetization of the surface cause a voltage pulse in the read head due to electromagnetic induction. Hard disks organize their data in so-called sectors (with e.g. 512, 2048 or 4096 bytes), which can only ever be read or written as a whole. Encryption of hard disks therefore usually takes place per sector.

2.9.2 CBC-AES Hard Disk Encryption

There are a large number of hard disk encryption software products on the market. Many of them use the **CBC-AES method**. Here, each sector is divided into blocks of 128 bits each and the blocks are encrypted one after the other using AES in CBC operating mode. This is generally considered sufficient for most security applications.

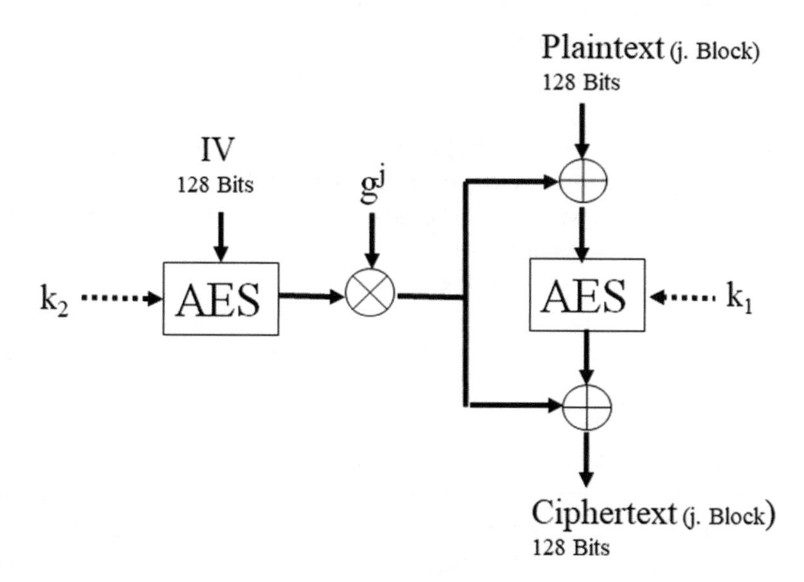

Fig. 2.14 XTS-AES hard disk encryption of the j. block

2.9.3 XTS-AES Hard Disk Encryption

However, both the **BSI** (Bundesamt für Sicherheit in der Informationstechnik, English: German Federal Office for Information Security) and NIST recommend in particular **XTS-AES** [BSI] for hard disk encryption. This is a standardized procedure that is also based on AES. The abbreviation **XTS** stands for "Xor-Encrypt-Xor-based tweaked-codebook mode with ciphertext stealing". With XTS-AES, too, each sector is divided into blocks of 128 bits, but AES is operated in an optimized ("tweaked") variant of the ECB mode. This is done with the following trick, which we have already seen for bytes (Sect. 2.8). Even for bit blocks of length 128, a reasonable addition $\oplus$ and multiplication $\otimes$ can be defined. For the 128-bit string t = 0…010, for example, the recursion formula $t^{128} = t^7 \oplus t^2 \oplus t^1 \oplus t^0$ can be used. Just like bytes, the bit blocks of length 128 then form a field in the mathematical sense. But we know more about a field, namely that there is at least one element g, which continuously exponentiated, i.e. $g^j = g \otimes \ldots_j \ldots \otimes g$ for j = 0, 1, 2,… yields all 128-bit blocks except 0…0 [Man, Wil, Buc]. Therefore, g is called a **generating element** (Sect. 2.5).

XTS-AES uses two AES keys. The key k_1 is used to AES-encrypt the 128-bit blocks per sector, and the other k_2 encrypts an initialization value IV of also 128 bits, which is usually derived from the sector address. The diagram in Fig. 2.14 schematically shows the work-flow of an XTS-AES encryption for the j. block within a sector. The procedure is as follows in detail:

- The initialization value IV with 128 bits is encrypted using AES and the key k_2.
- The result, again a string of 128 bits, is multiplied $\otimes$ by g^j.
- This string is added bitwise $\oplus$ to the plaintext of the j. block.
- The result is subjected to an AES cipher with key k_1.
- Finally, the string from the second point above is added bit by bit again $\oplus$.

The graphic in Fig. 2.14 serves as a simplified representation of the procedure. If one were to proceed in the same way in practice, the IV value would be encrypted again and again for each block of the sector and g^j would be recalculated again and again. This is unnecessary. Therefore, the encryption of IV per sector is done only once at the beginning, and g^j is calculated successively as $g^j = g^{j-1} \otimes g$.

If the division of the sector into blocks does not work, a rudimentary block of less than 128 bits remains at the end. This is then filled by the last bits of the ciphertext of the penultimate block ("ciphertext stealing") [WPDET].

Unlike CBC-AES, in the tweaked-ECB mode of XTS-AES each block is independent and not concatenated with other blocks. This means that if stored ciphered data is corrupted, only the data of that particular block is unrecoverable. However, XTS-AES requires AES keys twice as long, so 256 bits and 512 bits for AES-128 and AES-256 respectively.

Other storage media such as **USB sticks** (so-called flash memories) are also commercially available with CBC-AES or XTS-AES encryption [Kin].

2.9.4 ZIP Archive

The **ZIP** file format was originally developed in 1989 by **Phil Katz** (1962–2000). Today, there is a whole range of standard programs for creating and editing so-called ZIP archives, such as Winzip and 7-zip. The use of ZIP archives offers a whole range of advantages. They function as a container file into which several files belonging together or even entire directory trees can be packed. And they store data in compressed form, which was, incidentally, the real reason for their development. This way you can save space on your hard drive, fit more data on a USB stick, and upload and send it over the Internet is more practical. Incidentally, the compression method developed by Phil Katz is called **DEFLATE**.

Zip archives are also very popular because they can be optionally encrypted, which increases data security, especially when sending files. Encrypted ZIP archives can only be accessed by entering a password. The files of a ZIP archive are encrypted with DES in older versions, but with AES in newer versions, alternatively with the key lengths 128 bit or 256 bit.

3 Public-Key Ciphers

Up to now, all our encryption methods were designed in such a way that the encryption key was immediately known as the decryption key, or at least that it could be calculated without great difficulty. We called these methods symmetric ciphers (Sect. 2.1). In the case of asymmetric ciphers, it should be practically impossible to deduce the decryption key from the knowledge of the encryption key. Therefore, in this case, the encryption key can be made public. This is why these methods are also called **public-key ciphers**.

3.1 Factorization and RSA Cipher

3.1.1 Prime Numbers and Factorization

But what, on the one hand, should be easy to handle as a key, but, on the other hand, cannot be calculated in a reasonable amount of time, especially today, with our networked supercomputers? Mathematical topics probably come to mind, which have had great appeal since antiquity, but have steadfastly eluded a reasonable solution to this day. One of these problems is the decomposition of a natural number into factors, preferably prime numbers. A **prime number** is a natural number that is divisible only by 1 and itself, such as 2, 3, 5, 7, 11, 13, 17,… While it is known in principle that any natural number can be uniquely decomposed into its prime factors, e.g. $60 = 2^2 \cdot 3 \cdot 5$, how does one do this concretely? The obvious thing to do is to examine a given number for possible divisors. However, for very large natural numbers, this method quickly reaches its runtime limits. In short: **Factoring (in a reasonable time) is difficult**. However, one cannot prove this mathematically conclusively. In any case, if the puzzle is unexpectedly solved tomorrow, some of what we are about to learn will have to be completely rethought.

O. Manz, *Encrypt, Sign, Attack*, Mathematics Study Resources,
https://doi.org/10.1007/978-3-662-66015-7_3

3.1.2 Fermat's Little Theorem

How to translate the problem of factoring into a public-key cipher is what we want to look at now. The idea for this is based on the following statement, the so-called **Fermat's little theorem:** Let p be a prime number and a a natural number coprime with p. Then $a^{p-1} = 1$ (mod p) holds.

This statement, which goes back to **Pierre de Fermat** (1607–1665), contains again a modulo calculation (Sect. 1.2), as we already know it from letters (mod n = 26) or from bits (mod n = 2). If a = x (mod n) and b = y (mod n), then a • b = x • y (mod n) is also true. In words, this means, "Whether you first calculate the remainders modulo n and then multiply, or whether you first multiply and then calculate the remainders modulo n, it comes out to the same thing." This rule of calculation, which we shall use very frequently in what follows, is seen to be thus: Namely, if a = x + r • n and b = y + s • n with integers r and s, then a • b = (x + r • n)•(y + s • n) = x • y + (x • s + r • y + r • s • n) • n and therefore x • y = a • b (mod n). Thus, by our rule of arithmetic, if in particular $a^i = x$ (mod n) and $a^j = y$ (mod n), then $a^{i+j} = x \cdot y$ (mod n).

Let us also prove Fermat's little theorem for practice. First, we note that all numbers 1 • a, 2 • a,…, (p − 1) • a are distinct, and this is true even when considered as a remainder modulo p. Indeed, if i • a = j • a (mod p) holds with natural numbers i and j from the range 1 to p − 1, then (j − i) • a = 0 (mod p). But since the prime number p does not divide a by assumption, p must divide the difference j − i, and so j = i. Thus the numbers 1 • a, 2 • a,…, (p − 1) • a, each considered as a remainder modulo p, pass through all the remainders 1, 2,…, p − 1, but possibly in a different order. If we form their product in each case, then $a^{p-1} \cdot 1 \cdot 2 \cdot 3 \ldots \cdot (p-1) = 1 \cdot 2 \cdot 3 \ldots \cdot (p-1)$ (mod p) and p thus divides $(a^{p-1} - 1) \cdot 1 \cdot 2 \cdot 3 \ldots \cdot (p-1)$. Since p is a prime number, it must divide $a^{p-1} - 1$, so $a^{p-1} = 1$ (mod p).

3.1.3 Euclidean Algorithm

Before we come to the announced public key cipher, we want to remind you of the **Euclidean algorithm** [Wil, Buc]. It is named after **Euclid of Alexandria** (third century BC). The algorithm is used to determine the greatest common divisor of two natural numbers m and n. For this, let m be greater than n. Then set $r_0 = m$ and $r_1 = n$ and divide r_0 by r_1 with remainder, i.e. $r_0 = q_1 \cdot r_1 + r_2$ with r_2 less than r_1, and continue the procedure iteratively until at a k. step the division works out even:

$$
\begin{aligned}
r_0 &= q_1 \cdot r_1 + r_2 \\
r_1 &= q_2 \cdot r_2 + r_3 \\
\vdots &\quad \vdots \\
r_i &= q_{i+1} \cdot r_{i+1} + r_{i+2} \\
\vdots &\quad \vdots \\
r_{k-3} &= q_{k-2} \cdot r_{k-2} + r_{k-1} \\
r_{k-2} &= q_{k-1} \cdot r_{k-1} + r_k \\
r_{k-1} &= q_k \cdot r_k
\end{aligned}
$$

Then the Euclidean algorithm states that $g = r_k$ is the greatest common divisor of m and n. Calculating iteratively backwards from the next to last equation by substituting previous r_i, we get.

$$g = r_k = r_{k-2} - q_{k-1} \bullet r_{k-1} = r_{k-2} - q_{k-1} \bullet \left(r_{k-3} - q_{k-2} \bullet r_{k-2}\right) = \cdots$$

and the greatest common divisor g can finally be written as the multiple sum $g = x \bullet m + y \bullet n$ with integers x and y. These can be chosen so that x is positive and y is negative. Otherwise, namely, one modifies the multiple sum according to $g = x \bullet m + n \bullet m + y \bullet n - m \bullet n = (x + n) \bullet m + (y - m) \bullet n$. This is also called the **extended Euclidean algorithm** [Wil, Buc]. The method is highly efficient, fast and easy to implement.

3.1.4 RSA Cipher

The public key cipher we are about to describe was created by **Ronald Rivest** (b. 1947), **Adi Shamir** (b. 1952), and **Leonard Adleman** (b. 1945). It was published in 1977 and is known as the **RSA cipher**. How it works is shown schematically in Fig. 3.1 and described below.

The potential communication participant Y(ollanda) first obtains two different very large prime numbers p and q and multiplies them to the number $n = p \bullet q$. She also chooses a natural number e smaller than $(p - 1) \bullet (q - 1)$, which is coprime with $(p - 1) \bullet (q - 1)$.

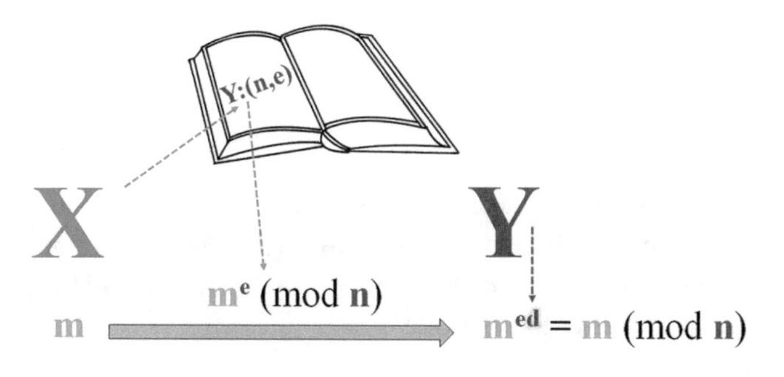

Fig. 3.1 RSA cipher

Using the extended Euclidean algorithm, she can then write the greatest common divisor 1 as the multiple sum of e and $(p - 1) \bullet (q - 1)$, that is, $1 = d \bullet e + b \bullet (p - 1) \bullet (q - 1)$ with a natural number d and a negative integer b. Our participant Y now registers in a central registry with her name and her so-called **public key** (n, e). She is then also said to have a **certified** RSA key. However, she keeps her **private key** d secret.

Now suppose sender X(avier) wants to send a secret message to receiver Y. X then first looks up the public key (n, e) of Y in the central register. Let the message m be a natural number smaller than the very large n. Now X sends the remainder of m^e when divided by n to Y. She uses the received m^e (mod n), takes her private key d and computes $(m^e)^d = m^{ed}$ (mod n). As a result, she receives $m = m^{ed}$ (mod n), and since m is smaller than n, exactly the desired plaintext m.

To see this, we first consider that the statement $m^{ed} = m^{1-b(p-1)(q-1)} = m \bullet (m^{(p-1)})^{-b(q-1)} = m$ (mod p) holds. Namely, if p is not a divisor of m, then Fermat's little theorem gives $m^{(p-1)} = 1$ (mod p). However, if p divides the number m, then both sides are equal to 0 modulo p. Similarly, for the other prime number q, $m^{ed} = m$ (mod q) also holds. Thus $n = p \bullet q$ is a divisor of $m^{ed} - m$, so $m^{ed} = m$ (mod n).

3.1.5 Example: RSA Cipher

(a) To explain the procedure concretely, we start with a very small example. Receiver Y(ollanda) chooses prime numbers p = 3 and q = 5. Therefore, n = 15 and $(p - 1) \bullet (q - 1) = 2 \bullet 4 = 8$. Since e = 3 is coprime with 8, she can choose (n, e) = (15, 3) as her public key. To determine her private key d, she uses the extended Euclidean algorithm for $(p - 1) \bullet (q - 1) = 8$ and e = 3. Here first is the iterated division with remainder.

$$\begin{aligned} 8 &= 2 \bullet 3 + 2 \\ 3 &= 1 \bullet 2 + 1 \\ 2 &= 2 \bullet 1 + 0 \end{aligned}$$

Since in the last equation the division works out even, the divisor 1 is the greatest common divisor, but this was already clear in this small example anyway. Much more important here is the fact that from the previous equations iteratively calculated backwards one can represent the greatest common divisor 1 as the multiple sum of $(p - 1) \bullet (q - 1) = 8$ and e = 3, viz.

$$1 = 3 - 1 \bullet 2 = 3 - 1 \bullet (8 - 2 \bullet 3) = 3 \bullet 3 - 1 \bullet 8 = 3 \bullet e - 1 \bullet (p-1) \bullet (q-1) = d \bullet e + b \bullet (p-1) \bullet (q-1).$$

So d = 3 is the private key of Y. For example, let m = 7 be the message to be sent. Then sender X(avier) computes the value m^e (mod n), so $7^3 = 343 = 13$ (mod 15), and therefore sends 13. Receiver Y uses her private key d = 3, computes $13^d = 13^3 = 2197 = 7$ (mod 15), and thus receives the message m = 7.

(b) Here is a slightly larger example. For the prime numbers p = 17 and q = 19, $n = 17 \cdot 19 = 323$ and $(p - 1) \cdot (q - 1) = 16 \cdot 18 = 2^5 \cdot 3^2 = 288$. Since e = 5 is coprime with 288, receiver Y(ollanda) can choose (n, e) = (323, 5) as her public key in this case. By means of the extended Euclidean algorithm we get

$$288 = 57 \cdot 5 + 3$$
$$5 = 1 \cdot 3 + 2$$
$$3 = 1 \cdot 2 + 1$$
$$2 = 2 \cdot 1 + 0$$

and consequently iteratively $1 = 3 - 1 \cdot 2 = 3 - 1 \cdot (5 - 1 \cdot 3) = 288 - 57 \cdot 5 - (5 - (288 - 57 \cdot 5)) = -115 \cdot 5 + 2 \cdot 288 = -115 \cdot 5 + 288 \cdot 5 - 3 \cdot 288 = 173 \cdot 5 - 3 \cdot 288 = d \cdot e + b \cdot (p - 1) \cdot (q - 1)$. This provides the private key d = 173 of receiver Y. If X(avier) wants to send the message m = 4 to Y, he computes $m^e = 4^5 = 55$ (mod n = 323) and sends 55. Receiver Y in turn computes $55^d = 55^{173} = 4$ (mod n = 323) and thus receives the message m = 4.

(c) Finally, a more complex example [Kob], also to show that one quickly reaches the limits of manual comprehension here. Let the prime numbers this time be p = 281 and q = 167, so $n = 281 \cdot 167 = 46{,}927$. Further, receiver Y(ollanda) chooses a random e = 39,423 such that e is coprime with $(p - 1) \cdot (q - 1) = 280 \cdot 166$. Then, using the extended Euclidean algorithm, she determines as her private key d = 26,767. Then, for example, if X(avier) wants to send the message m = 16,346, he must compute $m^e = 16{,}346^{39423}$ (mod n = 46,927), obtaining the remainder 21,166, which he sends to Y. Receiver Y thus computes $21{,}166^d = 21{,}166^{26767}$ (mod 46,927), gets 16,364 as the remainder, and so the original message m = 16,364.

3.1.6 Repeated Squaring

The RSA cipher requires a faster procedure for computing m^k (mod n) for natural numbers k than the obvious successive multiplication of m. To do this, write k as a binary expansion $k = k_{(r)} \cdot 2^r + \ldots + k_{(1)} \cdot 2^1 + k_{(0)} \cdot 2^0 = (\ldots((k_{(r)} \cdot 2 + k_{(r-1)}) \cdot 2 + k_{(r-2)}) \cdot 2 + \ldots + k_{(1)}) \cdot 2 + k_{(0)}$ with $k_{(i)} = 0$ or $k_{(i)} = 1$, but in any case $k_{(r)} = 1$. Then $m^k = ((\ldots((m^2 \cdot mk_{(r-1)})^2 \cdot mk_{(r-2)})^2\ldots)^2 \cdot mk_{(1)})^2 \cdot mk_{(0)}$. This is referred to as **repeated squaring**.

Each step consists of a square and, for $k_{(i)} = 1$, of an additional multiplication by m. If you do not want to calculate m^k itself, but only the remainder m^k (mod n), you form the remainder modulo n after each squaring and multiplication.

Let us illustrate this with an example [Hau1] and choose m = 296, k = 53 and $n = 13 \cdot 23 = 299$. Then $k = 53 = 2^5 + 2^4 + 2^2 + 2^0$ and therefore $m^k = 296^{53} = ((((296^2 \cdot 296)^2)^2 \cdot 296)^2)^2 \cdot 296$ We now calculate successively.

$$296^2 = (-3)^2 = 9 (\bmod 299)$$
$$296^2 \bullet 296 = 9 \bullet (-3) = -27 (\bmod 299)$$
$$(-27)^2 = 729 = 131 (\bmod 299)$$
$$(-27)^4 = 131^2 = 17161 = 118 (\bmod 299)$$
$$(-27)^4 \bullet 296 = 118 \bullet (-3) = -354 = 244 (\bmod 299)$$
$$244^2 = (-55)^2 = 3025 = 35 (\bmod 299)$$
$$244^4 = 35^2 = 29 (\bmod 299)$$
$$29 \bullet 296 = 8584 = 212 (\bmod 299)$$

Thus, $m^k = 296^{53} = 212$ (mod n = 299).

3.1.7 Security of the RSA Cipher

One can show the following [BNS, Kob]: Knowing the public key (n, e), it is just as "hard" to factorize n into its prime factors p and q as it is to compute the private key d. So this, if one believes in the statement "Factorizing in reasonable time is hard", is exactly the situation needed for a public key cipher. Nevertheless, it must also be clearly stated that it is unknown whether one really needs the private key d for decryption or whether there may be other efficient methods.

In order to attack RSA, one therefore tries as a matter of priority to develop factorization algorithms that are as fast as possible (Sect. 3.4). For security reasons, the BSI guideline [BSI1] recommends that the key length of the RSA module $n = p \bullet q$ should be of the order of 2000 bits as a binary expansion. Therefore, it is a natural number with about 600 decimal places. However, with increasing computer performance, the BSI recommends using RSA modules n with a length of 3000 bits for a deployment period beyond 2022. The other public key e can, however, be chosen to be quite small, but not too small for security reasons.

Among all possible attack methods (Sect. 2.1), chosen plaintext attacks are the most basic against public-key schemes, since any attacker knows the public keys and can therefore encrypt all plaintexts of his choice. Let us briefly consider here that RSA is vulnerable to certain chosen ciphertext attacks. To this end, let (n, e) be the RSA public key of participant X(avier). We assume attacker A(rchibald) intercepts the ciphertext c, but does not know the corresponding plaintext message m to $c = m^e$ (mod n). However, he cannot have it decrypted either, because that would be too conspicuous. Then he chooses a plaintext m_1 and encrypts it to $c_1 = (m_1)^e$ (mod n). Attacker A can assume that c_1 is coprime with n, otherwise he would have found a divisor of n and thus cracked the method. So he uses the extended Euclidean algorithm to compute the multiple sum representation $1 = x \bullet c_1 + y \bullet$

n. Consequently, $x \bullet c_1 = 1 \pmod n$, i.e., $(c_1)^{-1} = x \pmod n$. Thus he forms $c_2 = (c_1)^{-1} \bullet c \pmod n$ and lets the more innocuous c_2 decrypt to m_2, i.e. $c_2 = m_2^e \pmod n$. But then $c = c_1 \bullet c_2 = m_1^e \bullet m_2^e = (m_1 \bullet m_2)^e \pmod n$, so attacker A can decrypt the ciphertext c to the plaintext $m = m_1 \bullet m_2 \pmod n$.

3.1.8 RSA Cipher and Key Exchange

Given the size of n, it is not surprising that the RSA cipher is very slow despite the repeated squaring method, at least it requires much more computation time than symmetric ciphers. Therefore, both are usually used in combination in practice:

- With the public-key cipher RSA, one merely exchanges the key necessary for a symmetric cipher (so-called **key exchange**).
- For the actual transmission of information, the much faster symmetric cipher is then used, e.g. Triple-DES or AES with the secretly exchanged key.

This is also the reason why in practice the message m in an RSA cipher, i.e., the key of a symmetric cipher, is always smaller than the RSA module n. For the 128-bit key $k = k_0 k_1 k_2 \ldots k_{127}$ of AES, for example, one uses its binary expansion $m = k_0 \bullet 2^0 + k_1 \bullet 2^1 + k_2 \bullet 2^2 + k_3 \bullet 2^3 + \ldots + k_{127} \bullet 2^{127}$, to make it a natural number m for the RSA cipher.

3.2 Internet and WLAN

3.2.1 Network Protocols and the Internet

Computer networks require so-called **network protocols** to operate. A network protocol consists of a set of rules and formats that determine the communication behavior of the communicating instances within the networks. For systematization purposes, the so-called **OSI model** has become established, which divides the protocols into seven layers: from the lowest layer, pure bit transmission, to the highest layer, where the applications are

Table 3.1 OSI and TCP/IP layer model

Layers OSI		Layers TCP/IP	
7	Application	4	Application
6	Presentation		
5	Session		
4	Transport	3	Transport
3	Network	2	Internet
2	Link	1	Link
1	Physical		

located. The OSI model is shown in the left column of Table 3.1. The more rudimentary a communication is, the fewer layers are necessary or the more rudimentary the protocols on the upper layers can be. If, for example, a communication runs completely without users, no application layer is necessary. If only point-to-point connections are involved, no network layer is required.

Especially for the **Internet** and the Internet protocol family, the seven layers of the OSI model are usually combined into four levels in the **TCP/IP model** according to the right column of Table 3.1. The basic elements are the **IP** protocol (Internet Protocol), and **TCP** (Transmission Control Protocol), which organizes data transport. In the link layer, the **Ethernet** protocol is often used as well as **DSL** (Digital Subscriber Line) for fast bit transmission. The application layer is home to a variety of protocols.

3.2.2 Confidential Work on the Internet with HTTPS, SMTPS and FTPS

The protocol **TLS** (Transport Layer Security), formerly known as SSL (Secure Socket Layer), enables the secure transmission of information on the application layer via TCP/IP-based connections on the Internet. TLS is located at the upper end of the transport layer above TCP in the TCP/IP model, as Table 3.2 shows. It often works together with the following protocols of the application layer:

- **HTTP** (Hypertext Transfer Protocol), with which a user (client) can access the pages of a provider (server) by means of a browser.
- **SMTP** (Simple Mail Transfer Protocol), which is used to send e-mails.
- **FTP** (File Transfer Protocol), which allows files to be downloaded from a server to the client or uploaded from the client to the server.

To indicate the interaction with TLS, an S for "Secure" is usually appended to the protocol of the application layer, i.e. **HTTPS**, **SMTPS** and **FTPS**. If, for example, you call up an Internet page, you will find an "HTTPS" in its Internet address if the respective provider wants to particularly secure the content of this page. This is particularly the case if bookings can be made on the page or other transactions can be carried out. The short message service **Twitter** also uses the TLS infrastructure for secure data transmission.

Table 3.2 Protocol TLS in the TCP/IP model

Application	**HTTP, SMTP, FTP**
Transport	**TLS**
	TCP
Internet	IP
Link	Ethernet DSL

Preferably, TLS is operated with the symmetric cipher AES in CBC or CTR mode and keys of length 128 or 256. Triple DES no longer plays a significant role. RSA, among others, can be used as a public-key cipher. TLS consists of several subprotocols. With the **TLS Handshake Protocol**, the user (client) and provider (server) determine the cipher method to be used and agree on the key for the symmetric cipher. If RSA is used for the key exchange, the server sends its certified public RSA key to the client. The client then sends the server a secret random natural number encrypted with this key, which is to be used as the AES key. The server decrypts the random number with its private RSA key. After that, the **TLS Record Protocol**, the actual heart of TLS, can begin, which encrypts the communication on the Internet, for example, via browser with AES [WPTLS, WPTLSe, BSI2].

3.2.3 Wireless WLAN

WLAN (Wireless Local Area Network, or Wi-Fi) refers to a local wireless network. This can involve larger installations with a central server, but also in the private environment or in office communication, one likes to network devices (router, laptop, printer, etc.) wirelessly with each other with a WLAN. In connection with the Internet, it is often only used as an interface where you can dial into the Internet wirelessly with a laptop or smartphone via a nearby router.

As the successor to **WEP** (Wired Equivalent Privacy), which was considered insecure, the new WLAN standard specifies the **WPA2** (Wi-Fi Protected Access 2) method. WEP was based on the stream cipher **RC4** (Ron's Code 4) from 1987 by **Ronald Rivest** (born 1947), which was kept secret but was anonymously made public in 1994. WPA2, on the other hand, uses AES for data encryption with a key length of 128 bits in CTR operating mode.

In large WLAN installations, the server has a certified RSA key that can be used to exchange the AES key via the so-called **EAP-TLS protocol**. In smaller networks in the so-called SoHo domain (Small Office, Home Office), the **PSK** (Pre-shared key) procedure is usually used. The PSK key must be known to all devices in the WLAN, as it is used to generate the AES key. It can usually be entered on the various devices, and changing it regularly also increases security [WPWP2, BSI3].

3.3 Monte Carlo Prime Numbers

3.3.1 Prime Numbers for Public Key Ciphers

Now, at the latest, a fundamental question arises. How on earth is it possible to obtain sufficiently large prime numbers for an RSA cipher? One thing is reassuring: There are infinitely many prime numbers, hence arbitrarily large ones, as we have known since **Euclid**.

But in order to definitively determine whether a natural number p is really a prime number, one must actually check that it is not divisible by any number smaller than p, for which $\sqrt{p}$ is sufficient. In practice, however, it certainly cannot work like this, because otherwise one could search the number $n = p \bullet q$ for divisors in this way in a reasonable runtime. But that exactly this should not be possible was the basic idea of the RSA cipher. Is this already the practical end of RSA? Not quite: There are other **primality tests**. These are quite tricky, but usually have a flaw: They only prove with a certain probability that a given number is a prime number.

3.3.2 Carmichael Numbers

To understand the procedure, we start with a simple method based on Fermat's little theorem. Let n be an odd natural number, which we want to check whether it is a prime number. Let k be a fixed number of samples. Then we choose k random natural numbers a in the range from 2 to $n - 1$, which are coprime with n. The Euclidean algorithm can quickly decide on the coprimeness. If one finds an a for which a^{n-1} is not equal to 1 modulo n, then p is certainly not a prime number according to Fermat's little theorem. If one finds no such a after k samples, then n is at least possibly a prime number. Stupidly, it may actually happen that n passes this test even for all numbers a that are coprime with n, without actually being a prime number. Such numbers are called **Carmichael numbers,** and there are even an infinite number of them, the smallest being $561 = 3 \bullet 11 \bullet 17$. So this test yields the statement "possibly prime" for an infinite number of n, although this is factually not true at all.

3.3.3 Fermat's Theorem and 3. Binomial Formula

Let p be a prime number greater than 2 and let a be a natural number coprime with p. Then it follows from Fermat's little theorem and the 3. binomial formula that p divides the number $a^{p-1} - 1 = (a^{(p-1)/2} + 1)(a^{(p-1)/2} - 1)$. Since p is a prime number, p must divide either $a^{(p-1)/2} + 1$ or $a^{(p-1)/2} - 1$. If the second case is true, then because $a^{(p-1)/2} - 1 = (a^{(p-1)/4} + 1)(a^{(p-1)/4} - 1)$, it follows that p divides either $a^{(p-1)/4} + 1$ or $a^{(p-1)/4} - 1$. If here again the second case is true, then one continues this argument with $a^{(p-1)/8} + 1$ and $a^{(p-1)/8} - 1$. And so one can go on and on, as long as in each case the second case is true, and successively halve the exponent until this is no longer possible. This observation can also be expressed the other way round. For this, let $p - 1 = s \bullet 2^t$ with an odd number s. Then in our procedure at some point the first case is true, or p finally divides $a^s - 1$. In the first case, however, there is a j smaller than t in such a way that p is a divisor of $(\ldots((a^s)^2)^2\ldots_j\ldots)^2 + 1$.

3.3.4 Miller-Rabin Primality Test

The **Miller-Rabin primality test** dates from 1976 and is named after **Gary Miller** and **Michael Rabin** (b. 1931). It is based on the above simple corollary from Fermat's little theorem and works as follows: Let n be the odd natural number to be studied. For this, one again writes $n - 1 = s \bullet 2^t$ with an odd number s. Further, let k be a fixed number of samples. Then choose k random natural numbers a in the range from 2 to $n - 1$ that are coprime with n. The coprimeness is again easy to determine using the Euclidean algorithm. One checks then whether $a^s = 1 \pmod{n}$ or $a^s = -1 \pmod{n}$ is valid. If yes, then one makes a hook at the chosen sample a and takes the next random number. If no, then one checks by iterative squaring whether $(\ldots((a^s)^2)^2\ldots_j\ldots)^2 = -1 \pmod{n}$ holds for a j smaller than t. If this is the case, then one hooks the selected sample a and takes the next random number. If you cannot set a check mark for a sample, then n is certainly not a prime number. If, however, a check mark is placed on all samples, then n is a **prime number candidate** in the sense of Miller-Rabin.

After the experience with Carmichael numbers, one must now naturally ask the question: How many numbers pass the Miller-Rabin test as prime number candidates, even though they are not actually prime numbers? But now we are in a much better situation. You can show what is not so easy and therefore we will not do it here: The probability that a number n passes the Miller-Rabin test for a randomly chosen a, although it is not a prime number at all, is at most 1/4 [Kob, Buc]. Thus, if one performs the test for k independent samples a, the probability of error is at most $(1/4)^k$ and can therefore be made arbitrarily small by choosing enough samples. For k = 5, for example, this probability is already $(1/4)^5 = 1/1024$, i.e., smaller than 1‰. Natural numbers that pass the Miller-Rabin test can thus be safely used as prime numbers for an RSA cipher. This is indeed the way to get large prime numbers for RSA. The Miller-Rabin test is a so-called **Monte Carlo method**, i.e. a random-based method that only gives a false result with an upper bound probability.

3.3.5 Example: Miller-Rabin Primality Test

(a) Let us again consider an example [Hau3]. Let $n = 91 = 7 \bullet 13$, so $n - 1 = 45 \bullet 2$. Because of $3^{90} = 1 \pmod{91}$, $n = 91$ is possibly a prime number in the sense of Fermat with respect to the one sample $a = 3$.

We now perform the Miller-Rabin test with the one sample $b = 10$. The test is whether $b^{45} = 1 \pmod{91}$ or $b^{45} = -1 \pmod{91}$. Because of $10^{45} = -1 \pmod{91}$, $n = 91$ is a prime number candidate in the sense of Miller-Rabin with respect to the one sample $b = 10$. For the sample $a = 3$, however, $3^{45} = 27 \pmod{91}$ holds, so that the Miller-Rabin test with the sample $a = 3$ excludes the number $n = 91$ as a prime number.

(b) Here is yet another example [WPMRT]. We now want to test $n = 221 = 13 \bullet 17$ using the Miller-Rabin method. Because of $n - 1 = 220 = 55 \bullet 2^2$, $s = 55$ and $t = 2$. We choose as a random sample $a = 174$, which is coprime with $n = 221$, and calculate $a^s = 174^{55} = 47$

(mod n = 221). This is not equal to 1 and not equal to −1 modulo n = 221. So we also compute $(a^s)^2 = 47^2 = -1$ (mod n = 221). This shows that n = 221 is a prime number candidate in the sense of Miller-Rabin with respect to the one sample a = 174. We try a second sample b = 137. Then $b^s = 137^{55} = 188$ (mod n = 221), so again unequal 1 and unequal −1 modulo n = 221. Also $(b^s)^2 = 188^2 = 205$ (mod n = 221) is unequal −1 modulo n = 221. Therefore the sample b = 137 excludes the number n = 221 as prime number.

3.3.6 Euler Criterion

The so-called **Euler criterion**, named after **Leonhard Euler** (1707–1783), is a slight tightening of Fermat's little theorem, which we will not prove here [Wil] and which states the following: If p is a prime number greater than 2 and a is a natural number coprime with p, then

- $a^{(p-1)/2} = 1$ (mod p) if $a = b^2$ (mod p) is a square modulo p and
- $a^{(p-1)/2} = -1$ (mod p) for a non-square a modulo p.

Let us look at a simple example. Let p = 7. Then $1^2 = 1$ (mod 7), $2^2 = 4$ (mod 7), $3^2 = 2$ (mod 7), $4^2 = 2$ (mod 7), $5^2 = 4$ (mod 7), and $6^2 = 1$ (mod 7). So the squares modulo 7 are exactly the numbers 1, 2 and 4. For the square a = 2 (mod 7) we get $a^{(p-1)/2} = 2^3 = 1$ (mod 7), and for the non-square a = 3 (mod 7) we get $a^{(p-1)/2} = 3^3 = 6 = -1$ (mod 7) in agreement with Euler's criterion.

3.3.7 Solovay-Strassen Primality Test

Here is another primality test as an example of a Monte Carlo method, namely the **Solovay-Strassen primality test** published by **Robert Solovay** (b. 1938) and **Volker Strassen** (b. 1936) in 1977. It uses the Euler criterion and otherwise follows the same strategy as the Miller-Rabin test. Namely, let n be the odd natural number to be tested and k be a fixed number of samples. Then one chooses k random natural numbers a in the range from 2 to n − 1, which are coprime with n, and calculates $a^{(n-1)/2}$ (mod n). If one finds an a that does not satisfy the Euler criterion (for n instead of p), then n is certainly not a prime number. However, if the Euler criterion is satisfied for all samples a, then n is a **prime number candidate** in the sense of Solovay-Strassen.

If, in the Solovay-Strassen test, $a^{(n-1)/2} = 1$ (mod n) or = −1 (mod n), then one must also decide whether a is a square modulo n. To do this, one does not try to explicitly calculate a "square root" b modulo n from a, which would be difficult for large composite n anyway. Much more effective is the method of computing with so-called Legendre and Jacobi symbols [Wil]. However, the running time of the Solovay-Strassen test is still worse than that of Miller-Rabin. In addition, the probability that a number n passes the test for a, although it is not a prime number, is twice as high as in the Miller-Rabin test, namely at most 1/2 [Kob]. If, however, the test is carried out again for k independent samples a, the error probability is at most $(1/2)^k$, and here too we very quickly arrive at an extremely small residual risk.

3.3.8 AKS Primality Test

It is true that in 2002, with the **AKS primality test** [Wil, Hau3], a deterministic method was published for the first time, which thus identifies prime numbers as such with certainty and which also has "in principle a reasonable running time". However, despite some improvements in the meantime, this is still too high for practical applications.

3.4 Attack by Factorization

Unfortunately, if a primality test fails, one has no clue as to what factors this number has. But this is exactly what one would need to know in order to calculate the decomposition $n = p \bullet q$ into prime numbers from the public key (n, e) of an RSA cipher and thus crack the cipher. So, primality tests do not provide a starting point for cryptanalysis. Successively trying for divisibility for numbers up to $\sqrt{n}$ is much too slow for a magnitude of over 2000 bits, or 600 decimal places. So let's do some cryptanalysis again and look for faster methods for factorization in order to crack the RSA method or quantify its security.

3.4.1 Fermat Factorization

The following factorization method again goes back to **Pierre de Fermat** (1607–1665). The basic idea here is to write the natural number n to be factorized as the difference of two squares, i.e. $n = x^2 - y^2$ with two natural numbers x and y. Using the 3. binomial formula, this results in $n = x^2 - y^2 = (x + y) \bullet (x - y)$, i.e. a factorization of n.

But first you need the largest natural number s less than or equal to $\sqrt{n}$. This is best done with the **Heron method** according to **Heron of Alexandria**. You start with $x_0 = n$ and iteratively calculate $x_{i+1} = (x_i + n/x_i)/2$ until you find the first k such that $x_k - x_{k+1}$ is less than 1. Then s is the integer part of x_{k+1}. If $s = \sqrt{n}$, then already $n = s^2$. Otherwise, the **Fermat factorization** successively computes $(s + 1)^2 - n, (s + 2)^2 - n, \ldots, (s + i)^2 - n, \ldots$, and this until one finds a square number. Because of the differences relatively small numbers result, so that one can test this by successive trying, or one uses again the Heron method. However, one does not successively recalculate the squares $(s + i)^2$, but uses the 1. binomial formula $(s + (i + 1))^2 = ((s + i) + 1)^2 = (s + i)^2 + 2 \bullet (s + i) + 1$, thus adding $2 \bullet (s + i) + 1$ to the square of the predecessor $(s + i)^2$. Finally, if by this procedure one has identified an i for which $(s + i)^2 - n = a^2$ is a square number, then by the 3. binomial formula $n = (s + i)^2 - a^2 = (s + i + a) \bullet (s + i - a)$, and one has found two factors $s + i + a$ and $s + i - a$ of n.

The Fermat method thus searches for the divisor closest to $\sqrt{n}$ and arrives at a solution in a few iterations if the number n can be decomposed into two factors of approximately equal size. As a consequence for the security of the RSA cipher it follows that the two prime numbers p and q in $n = p \bullet q$ must not be too close to each other.

3.4.2 Example: Fermat Factorization

To illustrate, we give an example of Fermat factorization [Hau3, Kob] and choose n = 200,819. Some trial and error or Heron's method shows that the largest natural number s less than or equal to $\sqrt{200{,}819}$ is equal to 448. Then calculate $(s + 1)^2 - n = 449^2 - 200{,}819 = 782$. Unfortunately, this is not yet a square number, so we try s + 2 and now calculate $(s + 2)^2 - n = 450^2 - 200{,}819 = 1681$. This is a square number because of $1681 = 41^2 = a^2$, and we get $200{,}819 = n = (s + 2 + a) \bullet (s + 2 - a) = (450 + 41) \bullet (450 - 41) = 491 \bullet 409$.

On the other hand, if one chooses n = 141,467, s = 376, and it takes 38 steps with the Fermat factorization until one finally gets $n = 141{,}467 = 241 \bullet 587$.

3.4.3 Example and Basic Idea: Quadratic Sieve

Instead of determining natural numbers x and y with $n = x^2 - y^2$ for a given natural number n, one can also search for x and y more generally with $x^2 = y^2 \pmod{n}$. Namely, then n is a divisor of the difference $x^2 - y^2 = (x + y) \bullet (x - y)$, and if n does not divide x + y or x − y, then one can use the Euclidean algorithm to compute the greatest common divisor of n and x + y or x − y, respectively, and thus find a divisor of n. We make the procedure clear with an example [WPQSi]. Let n = 1649. Then for $x_1 = 41$, the equation $x_1^2 = 41^2 = 2^5 \pmod{1649}$, and for $x_2 = 43$, $x_2^2 = 43^2 = 2^3 \bullet 5^2 \pmod{1649}$ hold. If we multiply both equations, it follows that $(41 \bullet 43)^2 = 41^2 \bullet 43^2 = 2^5 \bullet 2^3 \bullet 5^2 = (2^4 \bullet 5)^2 \pmod{1649}$. Thus we have found $x = 41 \bullet 43$ and $y = 2^4 \bullet 5$ with $x^2 = y^2 \pmod{1649}$. Since the greatest common divisor of $x - y = 41 \bullet 43 - 2^4 \bullet 5 = 1683$ and n = 1649 equals 17, the desired factorization is $n = 1649 = 17 \bullet 97$.

In the search for x and y we therefore proceeded in such a way that we first determined the remainders y_i with $x_i^2 = y_i \pmod{n}$ for the samples x_1 and x_2. In our example, y_1 and y_2 had only the very small prime divisors 2 and 5. This construction principle can be used in general by allowing only small prime divisors $p_1 = 2, p_2 = 3, p_3 = 5, \ldots, p_{b-1}, p_b$ for y_i, up to a self-defined bound b. These prime numbers are then called **factor basis**. The factorization method originating from **John Dixon** thus seeks m distinct numbers x_i such that the remainders y_i of x_i^2 modulo n consist only of prime factors of the factor basis and that the remainder of $y_1 \bullet y_m$ modulo n is a square y^2. Indeed, the latter can then be easily read from the exponents of the factor basis [Wil]. In any case, from $x_i^2 = y_i \pmod n$ and $y_1 \bullet y_m = y^2 \pmod n$, it then follows for $x = x_1 \bullet x_m$, that $x^2 = y^2 \pmod n$.

Complementing Dixon's method, **Carl Pomerance** (b. 1944) has developed a method that systematically searches for the x_i using a sieve [Wil, Buc]. One therefore speaks altogether of the **quadratic sieve**.

3.4.4 Pollard's ρ- Factorization

We now want to get to know a method of factorization developed by **John Pollard** (born 1941), namely **Pollard's ρ-factorization** from 1975. Let n again be the natural number to be factorized. Then one considers a sequence x_i of natural numbers, starting for this purpose with a small natural number x_0, for example $x_0 = 1$ or $x_0 = 2$. One computes the sequence recursively by $x_{i+1} = x_i^2 + 1 \pmod n$ up to a x_b, imposing this "pain threshold" b on oneself and increasing it in case of failure. In the hope of finding a proper divisor of n, one now uses the Euclidean algorithm to successively determine the greatest common divisor of n and

$$x_1 - x_0$$

$$x_2 - x_0,\ x_2 - x_1$$

$$x_3 - x_0,\ x_3 - x_1,\ x_3 - x_2$$

$$\vdots \qquad \vdots \qquad \vdots$$

$$x_b - x_0,\ x_b - x_1,\ x_b - x_2,\ \ldots,\ x_b - x_{b-2},\ x_b - x_{b-1}$$

The name ρ-method is derived from the Greek letter ρ. One draws the sequence x_0, x_1, x_2,… as a chain of points. If one finds a prime number p as a divisor of n and $x_k - x_j$, then because of $x_k = x_j \pmod p$ the sequence of x_i becomes periodic modulo p, and the chain closes, so to speak, modulo p at the indices j and k. This is visualized in Fig. 3.2.

Pollard's ρ-method has a particularly good chance of finding a proper divisor of n if the number n has at least one smaller factor. As a consequence for the security of the RSA cipher it follows that the two prime numbers p and q in $n = p \bullet q$ should not be too far apart, because otherwise one of them becomes relatively small.

3.4.5 Example: Pollard's ρ-Factorization

We choose as an example [Hau3] the number n = 143, start with $x_0 = 2$ and choose as pain threshold b = 6. Then we get the following values for x_1 to x_6:

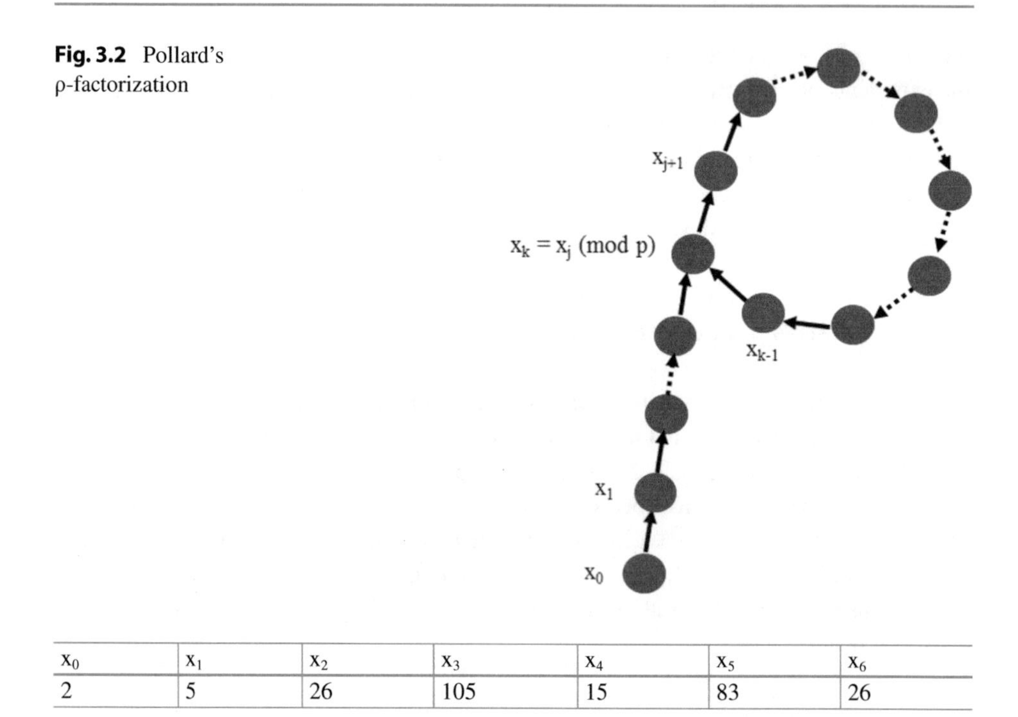

Fig. 3.2 Pollard's ρ-factorization

x_0	x_1	x_2	x_3	x_4	x_5	x_6
2	5	26	105	15	83	26

Now we successively compute the greatest common divisor of n = 143 and $x_k - x_j$ for k greater than j and k and j less than or equal to b = 6, finding the first proper divisor of n at $x_4 - x_0 = 13$. Thus 143 = 13 • 11.

3.4.6 Pollard's p − 1- Factorisation

The second factorization method by **John Pollard** (born 1941), which we will describe here, dates from 1974 and is called **Pollard's p − 1-factorization.** Let n again be the natural number to be factorized. First, one chooses a natural number b as the pain threshold and computes the least common multiple k of all the natural numbers from 1 to b. Then one chooses a random natural number a in the range from 2 to n − 1 and determines the greatest common divisor of a and n using the Euclidean algorithm. If you find a proper divisor, you have already factorized n. So, you can assume that a and n are coprime. Now calculate the remainder $c = a^k \pmod n$. If $c = 1 \pmod n$, then one tries another a or changes the pain threshold b. Otherwise, use the Euclidean algorithm to determine the greatest common divisor of c − 1 and n hoping to find a proper divisor of n.

When and why does Pollard's p − 1 method work? To examine this, suppose n has a prime divisor p such that p − 1 can be written as the product of relatively small prime powers; more precisely, all such powers should be at most equal to the pain threshold b. Then k is a multiple of p − 1, so we can write k as $k = (p - 1) \cdot k'$ with a natural number k′.

Because of $c = a^k \pmod{n}$, also $c = a^k \pmod{p}$, and from Fermat's little theorem follows $c = a^k = a^{(p-1)k'} = 1 \pmod{p}$, i.e., p divides both n and $c - 1$. So in this case one finds a proper divisor of n, unless n is itself a divisor of $c - 1$, i.e., unless $c = a^k = 1 \pmod{n}$ holds. But this was excluded above.

As a corollary for the security of the RSA cipher, it follows that for the two prime numbers p and q in $n = p \cdot q$, both $p - 1$ and $q - 1$ should have as large a prime divisor or prime power divisor as possible.

In this case Pollard's method reaches its limits. This is better done by using so-called **elliptic curves** (Sect. 3.7), as proposed by **Hendrik Lenstra** (b. 1949) [Wil].

3.4.7 Example: Pollard's p − 1 Factorization

Again, we consider an example [Kob]. Let n = 540,143. We choose a = 2 and b = 8, then k = 840 is the least common multiple of the numbers from 1 to 8. Because of $a^k = 2^{840} = 53{,}047$ (mod n = 540,143), c = 53,047, and the greatest common divisor of $c - 1 = 53{,}046$ and n = 540,143 calculates to 421 using the Euclidean algorithm. So $n = 421 \cdot 1283$.

An example [Kob] where the method works badly is $n = 491{,}389 = 383 \cdot 1283$ with prime numbers p = 383 and q = 1283, where $p - 1 = 2 \cdot 191$ and $q - 1 = 2 \cdot 641$ with prime numbers 191 and 641. Then b would already have to be at least 191 and k consequently a huge multiple of it for Pollard's p − 1 method to succeed.

3.5 Discrete Logarithm and Diffie-Hellman

We now want to take care of a second difficult problem of mathematics and see how to use it for a public key cipher. This is the so-called **discrete logarithm**, which is easier to understand than the term first suggests.

3.5.1 Existence of Generating Elements

Let p again be a prime number. We know from Fermat's little theorem (Sect. 3.1) that for all natural numbers c that are smaller than p and thus coprime with p, the statement $c^{p-1} = 1 \pmod{p}$ holds. What Fermat's little theorem does not rule out, however, is the possibility that there are also smaller exponents a in the range from 1 to $p - 2$ with $c^a = 1 \pmod{p}$. For example, for prime numbers p greater than 2, Euler's criterion (Sect. 3.3) says that for squares c modulo p, $c^{(p-1)/2} = 1 \pmod{p}$ already holds. Let us consider another concrete example. For p = 7 and c = 2, already $c^{(p-1)/2} = 2^3 = 8 = 1$ (mod p = 7). For g = 3, however, $3^1 = 3 \pmod{7}$, $3^2 = 2 \pmod{7}$, $3^3 = 6 \pmod{7}$, $3^4 = 4 \pmod{7}$, $3^5 = 5 \pmod{7}$ and $3^6 = 1 \pmod{7}$, and consequently here only g^{p-1} equals 1 modulo p.

This observation on a small example is also valid in general. For a prime number p greater than 2 there is always at least one natural number g less than p such that g^a is not equal to 1 modulo p for all a in the range from 1 to p – 2. One calls g a **generating element modulo p**. In contrast to Fermat's little theorem, however, this statement is not quite so obvious. We therefore omit a formal derivation [Wil, Buc].

The actual reason for the existence of generating elements is that the remainders modulo p form a field in the mathematical sense with respect to their addition and multiplication, as we have already seen with the bytes (Sect. 2.8) and with the example of hard disk encryption (Sect. 2.9). Every remainder r not equal to 0 modulo p has a multiplicative inverse. Since r and the prime number p are coprime, the inverse $r^{-1} = r_0 \pmod{p}$ is simply determined by the extended Euclidean algorithm to $r \bullet r_0 = 1 \pmod{p}$. It also follows that the powers $1 = g^0, g^1, g^2,\ldots, g^{p-2}$ of the generating element g pass through all residues not equal to 0 modulo p, thus **generating** them. Indeed, if $g^i = g^j \pmod{p}$, multiply by the inverse $g^{-i} = g^{p-1-i}$ modulo p and obtain $g^{(j-i)} = 1 \pmod{p}$. Consequently, i = j, and the g^i are pairwise different modulo p.

3.5.2 Determination of Generating Elements

The mathematical derivation that there is always at least one generating element g modulo a prime number p greater than 2 is only one side of the coin. But how to obtain such a generating element g_0 in concrete terms, especially when, as in our case, very large prime numbers p are involved, is a completely different question. The answer is: You do it by trial and error, so you choose a random natural number g_0 smaller than p and now you have to test in principle whether all g_0 powers g_0^a for a in the range from 1 to p – 2 are not equal to 1 modulo p. This will be very laborious for large p.

However, there is a criterion which makes the whole thing easier, but which we also do not want to derive formally [Wil, Buc]. Namely, one only has to prove that $g_0^{(p-1)/q}$ is not equal to 1 modulo p for all prime numbers q that divide p – 1. And the most efficient way to do this is to use the method of repeated squaring (Sect. 3.1). But what is the probability of success with the random choice of g_0? This in turn depends on the prime factorization of p – 1. If, for example, $p - 1 = 2 \bullet q$ with a prime number q, the probability is about 1/2.

We also consider a concrete example of this [Buc]. Let p = 23, so $p - 1 = 2 \bullet 11$. Then, $2^{11} = 3^{11} = 1 \pmod{23}$, and hence 2 and 3 are eliminated as generating elements modulo 23. However, $5^2 = 2 \pmod{23}$ and $5^{11} = 22 \pmod{23}$ as well as $7^2 = 3 \pmod{23}$ and $7^{11} = 22 \pmod{23}$ hold. Therefore, g = 5 as well as g = 7 are generating elements modulo 23.

3.5.3 Discrete Logarithm

Let a generating element g modulo the prime number p with p greater than 2 be given. For a given b the exponent a in $b = g^a \pmod{p}$ with a in the range from 0 to p – 2 is called the

discrete logarithm of b to the base g modulo p. The term discrete logarithm derives from the fact that this is the analogue of the logarithm of real analysis for finite (discrete) sets. For a large prime number p and given g and b, computing a by successively trying the powers g^1, g^2, g^3,... modulo p quickly reaches its limits. So we hold: **the discrete logarithm (in manageable time) is a difficult problem**. However, as with factoring, this cannot be proven mathematically conclusively.

We consider again a small example. For p = 19, we calculate that g = 2 is a generating element modulo 19. The discrete logarithm a of b = 7 to the base g = 2 is a = 6, because $2^6 = 64 = 7 \pmod{19}$.

3.5.4 Diffie-Hellman Key Exchange

As with the RSA cipher, the question now arises of how to convert this difficult mathematical problem into a practicable public-key cipher, which in principle can be used to send any message m in encrypted form (Sect. 3.8). However, we first recall the fact that public-key ciphers are usually only used in combination with symmetric ciphers for their key exchange. A method designed only for key exchange, based on the discrete logarithm, was

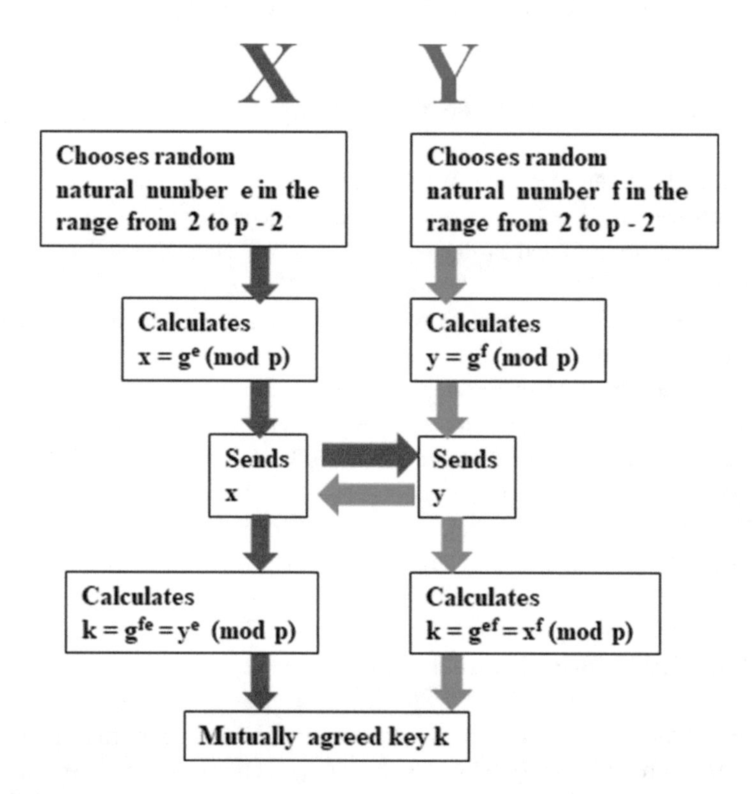

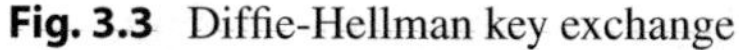
Fig. 3.3 Diffie-Hellman key exchange

published in 1976 by **Whit Diffie** (b. 1944) and **Martin Hellman** (b. 1945). It was these two who first proposed the idea of public-key cryptography in this seminal paper.

The **Diffie-Hellman key exchange** works as follows. Let p be a large prime number and g a generating element modulo p. Let both g and p be publicly known. We now imagine that X(avier) and Y(ollanda) want to agree on the key of a symmetric cipher for a planned secret data transmission. For this key exchange, they both use the following procedure, which is visualized in Fig. 3.3.

- X chooses a random natural number e in the range from 2 to $p - 2$.
- Y chooses a random natural number f in the range from 2 to $p - 2$.
- X sends the residue $x = g^e \pmod{p}$ to Y.
- Y sends the residue $y = g^f \pmod{p}$ to X.
- X calculates $k = g^{fe} = y^e \pmod{p}$
- Y calculates $k = g^{ef} = x^f \pmod{p}$.
- k is the mutually agreed key.

Attacker A(rchibald) could listen to x or y in the transmission channel. However, in order to be able to calculate $x^f \pmod{p}$ or $y^e \pmod{p}$, he would have to know e or f, i.e., he would have to solve the discrete logarithm for $x = g^e \pmod{p}$ or $y = g^f \pmod{p}$.

3.5.5 Example: Diffie-Hellman Key Exchange

(a) We start with a simple example [WPDHS]. We use the prime number $p = 13$ and convince ourselves that $g = 2$ is a generating element modulo $p = 13$. X(avier) chooses the random number $e = 5$, and Y(ollanda) chooses $f = 8$. Then X sends the value $6 = 2^5 \pmod{13}$ to Y, and Y sends $9 = 2^8 \pmod{13}$ to X. Now X computes the value $k = 3 = 9^5 \pmod{13}$, and Y computes $k = 3 = 6^8 \pmod{13}$. So $k = 3$ is the mutually agreed key for a symmetric cipher.

(b) Here is another slightly larger example [Kob]. We take as prime number $p = 53$ and the number $g = 2$, which is a generating element modulo $p = 53$. X(avier) chooses $e = 29$ and therefore transmits $45 = 2^{29} \pmod{53}$ to Y(ollanda). For her part, Y chooses $f = 19$ and therefore transmits $12 = 2^{19} \pmod{53}$ to X. Thereupon, X computes the number $k = 21 = 12^{29} \pmod{53}$ and Y computes the number $k = 21 = 45^{19} \pmod{53}$. After that, both X and Y know the common key $k = 21$.

3.5.6 Security of Diffie-Hellman Key Exchange

To crack the Diffie-Hellman method, the obvious thing to do is to try to compute the discrete logarithm a from $b = g^a \pmod{p}$. So this, if one believes in the statement "discrete logarithm in reasonable time is hard", is again exactly the situation needed for a public-key

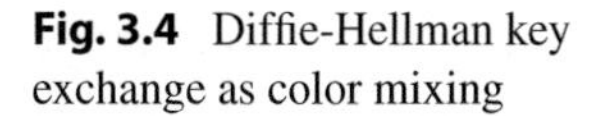

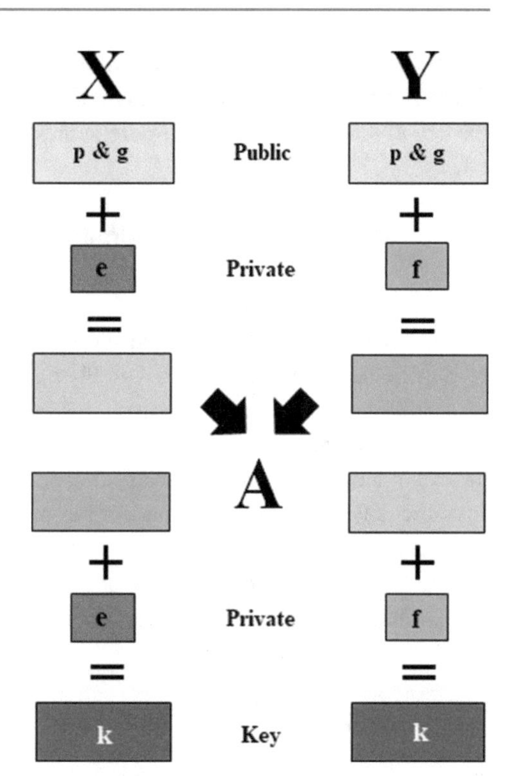

Fig. 3.4 Diffie-Hellman key exchange as color mixing

cipher. However, it is unclear whether one can break the Diffie-Hellman method using only the discrete logarithm, or whether there may be entirely different efficient methods.

In order to attack Diffie-Hellman, one therefore primarily tries to develop fast algorithms for calculating the discrete logarithm (Sect. 3.6). For security reasons, the BSI guideline [BSI1] recommends that the key length of the prime number p should be of the order of 2000 bits. These are therefore prime numbers with about 600 decimal places. We have already examined how to obtain such prime numbers using the Monte Carlo method for the RSA procedure (Sect. 3.3). However, with increasing computer performance, the BSI recommends using prime numbers p with a length of 3000 bits for a period of use beyond 2022.

3.5.7 Diffie-Hellman in Colour

One can also think of the Diffie-Hellman key exchange somewhat more vividly as a color game, as illustrated in Fig. 3.4. X(avier) and Y(ollanda) know the common color yellow (namely p and g), which every potential attacker A(rchibald) also knows. They each mix yellow with a secret color red (namely e) and turquoise (namely f) known only to them. These mixed colors brown and blue are now exchanged, and it is assumed that they are observed doing so. The underlying assumption is that A will not be able to determine exactly the red and turquoise used from the mixed colors. But this would be necessary in order, as X and Y do, to subsequently produce the common military green mixture (namely the key k) that they use for their communication.

3.5.8 Semi-Static Diffie-Hellman

Diffie-Hellman key exchange is also sometimes used as a so-called **semi-static** procedure, as follows. One of the two, say participant Y(ollanda), chooses a prime number p, a generating element g modulo p, and a number a, and computes $b = g^a \pmod p$. She publishes (p, g, b) as her public key and keeps a secret as her private key. If she wants to agree on a key k for a symmetric cipher with participant X(avier), she "statically" uses her secret a instead of a random number f and also does not need to send $y = b = g^a = g^f \pmod p$ to X anymore, since b is known as part of her public key anyway. X, on the other hand, still chooses a random number e, sends $g^e \pmod p$ to Y, and computes $k = g^{fe} = b^e$ (mod). Y uses her private key a and computes $k = g^{ef} = g^{ea} \pmod p$, and thus the shared key k is agreed upon. Thus, in this procedure, Y has a certified Diffie-Hellman key, but X does not.

3.6 Attack with Baby and Giant Steps

3.6.1 Baby-Step-Giant-Step Method

So let's do a little cryptanalysis again. In a brute-force attack on Diffie-Hellman, one would have to calculate all powers of the generating element g modulo p by successive multiplication until one finds the desired a with $b = g^a \pmod p$. So this method needs p multiplications in order of magnitude. With a little trick this can be done much faster. We take the smallest natural number t greater than or equal to $\sqrt{(p-1)}$ (Sect. 3.4) and formally divide the unknown a, which we can assume to be less than p – 1, by t with remainder. So $a = t \cdot q + r$ with natural numbers q and r, where r is less than t, holds. Then $b = g^a = g^{tq+r} \pmod p$ can also be written as $g^r = b \cdot g^{-tq} \pmod p$.

The **baby-step-giant-step method** from 1971 by **Daniel Shanks** (1917–1996) searches for two numbers q and r with this property. To do this, one first creates a table of baby steps consisting of r and $g^r \pmod p$ for r in the range from 0 to t – 1. Then, one successively computes the giant steps $b \cdot g^{-tq} \pmod p$, starting from q equal to 0 to at most t – 1, comparing the values with those in the baby step table. As soon as a match is found, the corresponding numbers r and q yield the sought $a = t \cdot q + r$.

The complexity of Baby-Step-Giant-Step results mainly from the successive multiplications modulo p. Since q and r are smaller than $\sqrt{p}$, one obtains a total order of $\sqrt{p}$ multiplications and thus a significant speedup compared to the brute-force attack. However, baby-step-giant-step also requires significantly more memory.

3.6.2 Example: Baby-Step-Giant-Step Method

We consider an example [WPBSG]. We choose as prime number $p = 29$. Then we can convince ourselves that $g = 11$ is a generating element modulo $p = 29$. Moreover, $t = 6$. We look for the discrete logarithm to the base $g = 11$ modulo $p = 29$ for $b = 3$, that is, the a with $3 = 11^a \pmod{29}$. Then, as a baby step table, we get.

r	0	1	2	3	4	5	
11^r	1	11	5	26	25	14	(mod 29)

Because of $g^{-t} = 11^{-6} = 11^{28-6} = 11^{22} = 13 \pmod{29}$, the giant steps are successively calculated as $3 \bullet 13^q$, and we obtain.

q	0	1	2	
$3 \bullet 13^q$	3	10	14	(mod 29)

At 14 (mod 29), there is a match with the baby-step table. Therefore $r = 5$, $q = 2$ and thus the sought-after a equals $a = t \bullet q + r = 6 \bullet 2 + 5 = 17$.

3.6.3 Pohlig-Hellman Method

The **Pohlig-Hellman method** by **Stephen Pohlig** (1953–2017) and **Martin Hellman** (b. 1945) was published in 1978. It also allows the computation of the discrete logarithm a of $b = g^a \pmod{p}$ for a prime number p and a generating element g modulo p. The method is faster than baby-step-giant-step when all prime divisors q of $p - 1$ are quite small. The trick works like this: To a prime q, let q^e be the highest q power that divides $p - 1$. Then one first determines the remainder c with $c = a \pmod{q^e}$ for initially still unknown a.

If this is done for all prime divisors q of $p - 1$, then there is a well-known standard technique to determine the sought a smaller than $p - 1$. This is the so-called **Chinese remainder theorem** [Wil, Buc], which already appears in various early Chinese sources, probably for the first time around 300 A.D. For $i = 1,\ldots, s$ we denote by n_i the highest q_i -power dividing $p - 1$, and set $n_i' = (p - 1)/n_i$. Since n_i and n_i' are coprime for each i, we use the extended Euclidean algorithm to find natural numbers x_i with $x_i \bullet n_i' = 1 \pmod{n_i}$, i.e. $(n_i')^{-1} = x_i \pmod{n_i}$. Since, according to the assumption, we have already determined all c_i with $c_i = a \pmod{n_i}$, we can compute $y_i = c_i \bullet (n_i')^{-1} \pmod{n_i}$ as well as $x = y_1 \bullet n_1' + \ldots + y_s \bullet n_s'$. Since each n_i is a divisor of n_j' for different i and j, we obtain $x = y_i \bullet n_i' = c_i \bullet (n_i')^{-1} \bullet n_i' = c_i = a \pmod{n_i}$. Because of $n_1 \bullet n_s = p - 1$ with pairwise coprime n_i, it follows that $a = x \pmod{p - 1}$.

So it remains to calculate c with $c = a \pmod{q^e}$. For this, first think of $c = c_{(0)} \bullet q^0 + c_{(1)} \bullet q^1 + c_{(2)} \bullet q^2 + \ldots + c_{(e-1)} \bullet q^{e-1}$ written as an expansion of q-powers with coefficients $c_{(i)}$ smaller than q, as we have already done for binary expansions (i.e. for q = 2). Let $h = g^{(p-1)/q} \pmod p$, then $h^q = 1 \pmod p$, and because $c = a \pmod{q^e}$ it follows $b^{(p-1)/q} = g^{a(p-1)/q} = h^a = h^c = h^{c_{(0)}} \pmod p$. Since $c_{(0)}$ is less than q, one can determine from it $c_{(0)}$ by successively comparing $b^{(p-1)/q} \pmod p$ with $1, h, h^2, \ldots, h^{q-1} \pmod p$. This is effective since q should be quite small by assumption. Now consider $b_1 = b \cdot g^{-c_{(0)}} = g^{a-c_{(0)}} \pmod p$. Then $b_1^{(p-1)/(q^2)} = g^{(a-c_{(0)})(p-1)/(q^2)} = h^{(a-c_{(0)})/q} = h^{(c-c_{(0)})/q} = h^{c_{(1)}} \pmod p$, and because $c_{(1)}$ is smaller than q, one can in turn determine $c_{(1)}$ from this by successively comparing $b_1^{(p-1)/(q^2)} \pmod p$ with $1, h, h^2, \ldots, h^{q-1} \pmod p$. Next, one forms $b_2 = b \bullet g^{-c_{(0)} - c_{(1)}q} = g^{a-c_{(0)}-c_{(1)}q} \pmod p$ and $b_2^{(p-1)/(q^3)} = h^{c_{(2)}} \pmod p$, determines from that $c_{(2)}$ and calculates in this way successively all coefficients $c_{(i)}$ and thus also c itself.

3.6.4 Example: Pohlig-Hellman Method

We will also give an example [Wil], which must be sufficiently complicated to demonstrate the Pohlig-Hellman method comprehensively. We choose the prime number p = 1999, thus $p - 1 = 1998 = 2 \bullet 3^3 \bullet 37$. As one has to verify, g = 3 is a generating element modulo p = 1999. For b = 1996 we want to determine the discrete logarithm a to the base g = 3 modulo p = 1999, for which thus $1996 = b = g^a = 3^a \pmod{p = 1999}$ holds.

We first examine the three prime divisors 2, 3, and 37 of p – 1. Using the terms of the Pohlig-Hellman method, we start with q = 2 and calculate $c = a \pmod 2$. To do this, we first note that $c = c_{(0)}$ holds. Because of $h = g^{(p-1)/q} = 3^{1998/2} = -1 \pmod{1999}$ and $b^{(p-1)/q} = 1996^{1998/2} = 1 \pmod{1999}$ it follows that $c_{(0)} = 0$ and therefore c = 0.

We next consider q = 3 and now calculate $c = a \pmod{3^3}$. To do this, we again note that $c = c_{(0)} + c_{(1)} \bullet 3 + c_{(2)} \bullet 3^2$ holds. Because of $h = g^{(p-1)/q} = 3^{1998/3} = 808 \pmod{1999}$ and $b^{(p-1)/q} = 1996^{1998/3} = 808 \pmod{1999}$, it first follows $c_{(0)} = 1$. Because of $b_1 = b \bullet g^{-c_{(0)}} = b \bullet g^{-1} = 1999 \bullet 3^{-1} = -1 \pmod{1999}$ and $b_1^{(p-1)/(q^2)} = -1^{(1998/9)} = 1 \pmod{1999}$, it now follows $c_{(1)} = 0$. Finally, $b_2 = b \bullet g^{-c_{(0)} - c_{(1)}q} = b \bullet g^{-1} = b_1 = -1 \pmod{1999}$, so that $b_2^{(p-1)/(q^3)} = -1^{(1998/27)} = 1 \pmod{1999}$ finally yields $c_{(2)} = 0$. Altogether, therefore, $c = c_{(0)} = 1$.

It remains to consider q = 37. We again look for $c = a \pmod{37}$ and note that $c = c_{(0)}$ holds. Now $h = g^{(p-1)/q} = 3^{1998/37} = 1309 \pmod{1999}$ as well as $b^{(p-1)/q} = 1996^{1998/37} = 1309 \pmod{1999}$, from which follows $c_{(0)} = 1$ and thus also c = 1.

Now, using the notation from the Pohlig-Hellman method, we collect what we already know in our example. It is $n_1 = 2$ and $n_1' = 3^3 \bullet 37$, $n_2 = 3^3$ and $n_2' = 2 \bullet 37$, and $n_3 = 37$ and $n_3' = 2 \bullet 3^3$. Also, we have just determined $c_1 = 0$, $c_2 = 1$, and $c_3 = 1$. Then the Chinese remainder theorem states that the sought a can be calculated as $a = x \pmod{1998}$ with x =

$y_1 \bullet n_1' + y_2 \bullet n_2' + y_3 \bullet n_3' = y_1 \bullet 3^3 \bullet 37 + y_2 \bullet 2 \bullet 37 + y_3 \bullet 2 \bullet 3^3$. Here $y_1 = c_1 \bullet (n_1')^{-1} = 0 \pmod 2$, $y_2 = c_2 \bullet (n_2')^{-1} = (2 \bullet 37)^{-1} \pmod{3^3}$ and $y_3 = c_3 \bullet (n_3')^{-1} = (2 \bullet 3^3)^{-1} \pmod{37}$. Using the extended Euclidean algorithm, one now determines y_2 and y_3 according to $y_2 \bullet 2 \bullet 37 = 1 \pmod{3^3}$ and $y_3 \bullet 2 \bullet 3^3 = 1 \pmod{37}$. Altogether, $y_2 = 23$ and $y_3 = 24$, hence $y_2 \bullet 2 \bullet 37 = 1702$ and $y_3 \bullet 2 \bullet 3^3 = 1296$ and hence $a = x = 1702 + 1296 = 1000 \pmod{1998}$.

3.6.5 Pollard's ρ-Method for Discrete Logarithms

We also want to discuss a second ρ-method by **John Pollard** (born 1941), namely the one for discrete logarithms from the year 1978. Thus, we are again looking for a with $b = g^a \pmod p$ for a prime number p and a generating element g modulo p. To do this, one first divides the numbers 1, 2,…, p − 1 into three roughly equal ranges B_1, B_2 and B_3, say B_1 from 1 to an n_1, B_2 from $n_1 + 1$ to an n_2 and B_3 from $n_2 + 1$ to p − 1. Inspired by Pollard's ρ-factorization, one now defines a sequence x_i of natural numbers in the range from 1 to p − 1 and starts for this with a random number $x_0 = g^{k_0} \pmod p$. The sequence itself is calculated recursively by

$$\begin{aligned} x_{i+1} &= g \bullet x_i \pmod p, && \text{if } x_i \text{ lies in the range } B_1 \\ x_{i+1} &= x_i^2 \pmod p, && \text{if } x_i \text{ lies in the range } B_2 \\ x_{i+1} &= b \bullet x_i \pmod p, && \text{if } x_i \text{ lies in the range } B_3 \end{aligned}$$

Let $m_0 = 0$. Then we can convince ourselves that x_i can be written as $x_i = g^{k_i} \bullet b^{m_i} \pmod p$ with

$$\begin{aligned} k_{i+1} &= k_i + 1 \pmod{p-1} && \text{and} \quad m_{i+1} = m_i, && \text{if } x_i \text{ lies in the range } B_1, \\ k_{i+1} &= 2 \bullet k_i \pmod{p-1} && \text{and} \quad m_{i+1} = 2 \bullet k_i \pmod{p-1}, && \text{if } x_i \text{ lies in the range } B_2, \\ k_{i+1} &= k_i && \text{and} \quad m_{i+1} = m_i + 1 \pmod{p-1}, && \text{if } x_i \text{ lies in the range } B_3, \end{aligned}$$

But since x_i can take only the finitely many values 1, 2,…, p − 1, there must be among the k_i and m_i numbers k and k′ as well as m and m′ with $g^k \bullet b^m = g^{k'} \bullet b^{m'} \pmod p$ and the chain closes, so to speak, modulo p. Putting here $b = g^a \pmod p$ and summing up the exponents of g, it follows for these $k - k' = a \bullet (m' - m) \pmod{p-1}$. In order to be able to calculate the discrete logarithm a concretely, one must first get the values k, k′, m and m′ and afterwards determine the sought a from the just derived equation modulo p − 1. If there are several solutions a, one must determine the correct one by trial and error or alternatively start with a new $x_0 = g^{k_0}$.

3.6.6 Example: Pollard's ρ-Method for Discrete Logarithms

As an example [WPPRL] for Pollard's ρ-method we choose the prime number p = 1019 and as generating element modulo p = 1019 the number g = 2, which of course has to be checked again. We look for the discrete logarithm a for b = 5 to the base g = 2 modulo p = 1019, so $5 = 2^a \pmod{1019}$. A small computer program helps to calculate $2^{681} \bullet 5^{378} = 1010 = 2^{301} \bullet 5^{416} \pmod{1019}$, and consequently $a \bullet (416 - 378) = (681 - 301) \pmod{1018}$. In any case, for this a = 10 is a solution with $2^a = 2^{10} = 1024 = 5 \pmod{1019}$. There is also a second solution a′ = 519, but for which $2^{a'} = 2^{519} = 1014 = -5 \pmod{1019}$ holds.

We will not discuss here another method for computing discrete logarithms, the **index-calculus method** [Wil, Buc].

3.7 Bluetooth and ECDH

3.7.1 Bluetooth Radio Interface

As a practical example, we will now look at the **Bluetooth** radio interface. This is a standard developed in the 1990s by the **Bluetooth Special Interest Group** for data transmission over short distances via radio technology. The devices involved transmit in a license-free so-called **ISM band** (Industrial, Scientific, Medical Band) at about 2.4 GHz and may be operated worldwide without approval. Depending on the transmission power, the range is between 1 m and 100 m, whereby the characteristics of the environment, such as the presence of partitions, can also strongly influence the range. The name Bluetooth is derived from the Danish king **Harald Blauzahn.** The main purpose of Bluetooth is to replace cable connections between different devices. Bluetooth provides an interface through which small mobile devices such as mobile phones and tablets as well as computers, printers and other peripheral devices can communicate with each other and exchange data.

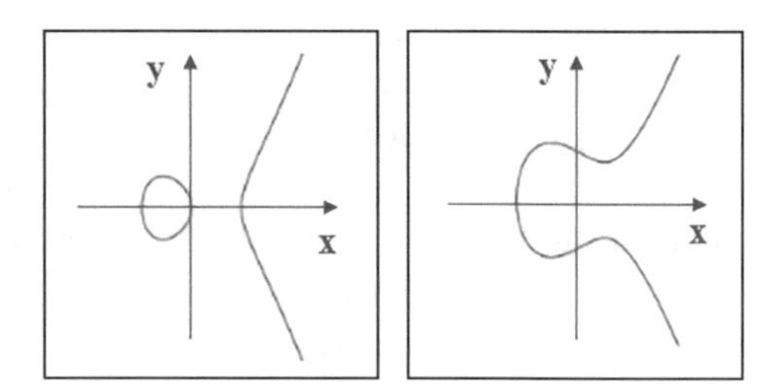

Fig. 3.5 Graphs of the elliptic curves $y^2 = x^3 - x$ and $y^2 = x^3 - x + 1$ in the real (x, y) plane (schematic)

3.7.2 Bluetooth Data Encryption

Like any radio transmission, Bluetooth offers potential attackers good opportunities to hack into the communication, especially at greater distances. For this reason, Bluetooth data are transmitted in encrypted form. The original encryption method was very similar to that used for GSM mobile radio (Sect. 2.3). The so-called E2 procedure and a random natural number RAND were used to generate the key for the actual encryption method E0. E0 was a stream cipher in which the pseudo-random sequences were generated with the aid of four shift registers of lengths 25, 31, 33 and 39 [Fox].

In the meantime, however, Bluetooth has switched to newer methods. For the encryption of the data to be transmitted, AES with a key length of 128 bits is used in CRT operating mode. The key exchange for the AES key takes place with the Diffie-Hellman method [Blu].

3.7.3 Elliptic Curves Modulo a Prime Number

But this was not yet the full truth. In order to explain this, however, we must first venture a small digression. In real analysis, **elliptic curves** are functions of the form $y^2 = x^3 + r \cdot x + s$ with real coefficients r and s and variables x and y (together with the technical condition that $4 \cdot r^3 + 27 \cdot s^2$ does not equal 0). For example, $y^2 = x^3 + 5 \cdot x + 3$ is an elliptic curve, and the point $P = (x_0, y_0) = (1, 3)$ lies on the curve since $y_0^2 = 9 = 1^2 + 5 \cdot 1^2 + 3 = x_0^3 + 5 \cdot x_0 + 3$ holds. Figure 3.5 shows the graphs of two typical elliptic curves in the real (x, y) plane.

Since one can add and multiply remainders modulo a prime number p, an elliptic curve is also conceivable modulo p. This means that the coefficients r and s are remainders modulo p and that also for the variables x and y only remainders modulo p are admissible. At first this sounds like abstract mathematical gimmickry, especially since one cannot imagine and plot the graph with its finitely many points in such a concrete way. However, one can operate with two points $P = (x_0, y_0)$ and $Q = (x_1, y_1)$ of the elliptic curve with residues x_0, x_1, y_0 and y_1 modulo p in such a way that the result is again a point on the curve, thus satisfying the curve equation. This operation is geometrically motivated from the real (x, y)-plane [BNS, Kob], but it is quite complicated. For the sake of simplicity it is written as addition P + Q, although it has nothing at all to do with a simple addition $x_0 + x_1$ and $y_0 + y_1$ of the coordinates.

3.7.4 Addition of Points on Elliptic Curves

So here are the calculation rules for the addition of two points, but only for information or even for "deterrence", and explicitly with the possibility to "skim" or even skip this. For further understanding, the intuitive notion of an additive operation of two curve points is quite sufficient. Let $P = (x_0, y_0)$ and $Q = (x_1, y_1)$ be two points on an elliptic curve $y^2 = x^3 + r \cdot x + s$ modulo a prime number p, where we want to assume that p is greater than 3.

First, define a fictitious point O of the curve, the so-called point at infinity, which is to be the neutral element of the addition, i.e. O + P = P + O = P. Moreover, we set $-P = (x_0, -y_0)$ and $P + (-P) = O$. We now want to define in general how to compute the sum

$P + Q = (x_S, y_S)$ with remainders x_S and y_S modulo p. Since we have already done this for $Q = -P$, we can assume that Q is not equal to −P.

If in addition Q is unequal to P, then x_0 is also unequal to x_1. Otherwise from the formula for the elliptic curve $0 = y_1^2 - y_0^2 = (y_1 - y_0) \bullet (y_1 + y_0) \pmod p$ would follow and therefore $y_1 = y_0$ or $y_1 = -y_0$. Thus one can divide by $x_1 - x_0$ modulo p (Sect. 3.5). One then defines

$$x_S = ((y_1 - y_0)/(x_1 - x_0))^2 - x_0 - x_1 \pmod p$$
$$y_S = -y_0 + ((y_1 - y_0)/(x_1 - x_0)) \bullet (x_0 - x_S) \pmod p$$

In the remaining case Q = P it holds that $y_0 = y_1$ is not equal to 0, because otherwise $Q = -P$. Therefore one can divide by y_0 modulo p, and one finally defines.

$$x_S = ((3 \bullet x_0^2 + r)/(2 \bullet y_0))^2 - 2 \bullet x_0 \pmod p$$
$$y_S = -y_0 + ((3 \bullet x_0^2 + r)/(2 \bullet y_0)) \bullet (x_0 - x_S) \pmod p$$

First you have to prove that P + Q is again a point on the elliptic curve, i.e. that the equation of the curve $y_S^2 = x_S^3 + r \bullet x_S + s$ is fulfilled. If you are motivated enough, you can also try to prove that the addition of the points satisfies reasonable calculation rules, i.e. P + Q = Q + P (commutative law) as well as (P + Q) + R = P + (Q + R) (associative law) for a third point $R = (x_2, y_2)$ on the elliptic curve.

3.7.5 Base Points on Elliptic Curves

So, at the latest now, elliptic curves modulo p become interesting also for cryptography. The idea is this: Let one choose a large prime number p and an elliptic curve $y^2 = x^3 + r \bullet x + s$ modulo p, by fixing its coefficients r and s. Also, find a point G on the curve where the iterative addition $i \bullet G = G + \ldots_i\ldots + G$ passes through as many different points on the curve as possible. One calls the number o after which a point repeats for the first time in this process, i.e., $j \bullet G = o \bullet G$ for a suitable j, the **order** o of G. But this in turn implies $O = (o - j) \bullet G$, and by the choice of o, j = 0 and $o \bullet G = O$ is the neutral element of the addition.

From experience one knows, or at least believes to know, that it is a hard problem to determine the natural number n for a **base point** G of large order o from the knowledge of a point $P = n \bullet G$. One suspects that this is even significantly more difficult than determining the discrete logarithm a of $b = g^a \pmod p$. So in this comparison the base point G plays the role of the generating element g modulo p. The baby-step-giant-step method, the Pohlig-Hellman method, and Pollard's ρ-method (Sect. 3.6) are suitably "translated" also applicable to calculate the number n from $P = n \bullet G$. But all known methods are much less efficient for elliptic curves than for the discrete logarithm.

3.7.6 ECDH

We now want to formulate the Diffie-Hellman key exchange for elliptic curves **ECDH** (Elliptic Curve Diffie Hellman). Let the values p, r, s, G and o be defined as above and publicly known.

- X chooses a random natural number e in the range from 2 to o − 1.
- Y chooses a random natural number f in the range from 2 to o − 1.
- X sends e • G to Y.
- Y sends f • G to X.
- X calculates P = e • f • G.
- Y calculates P = f • e • G.
- The x-component of P is the mutually agreed key.

An unauthorized person eavesdropping on e • G and f • G would have to be able to compute e or f efficiently in order to crack the procedure. The use of elliptic curves in cryptography was proposed by **Neal Koblitz** (born 1948) in 1987.

Like the normal Diffie-Hellman method, ECDH can also be used as a semi-static key exchange for a symmetric cipher (Sect. 3.5). In this case, only one of the two communication partners, say participant Y(ollanda), has a certified ECDH key, namely the public key (p, r, s, G, o, B) with B = b • G and the private key f = b. The other communication partner X(avier) continues to operate with a random number e.

3.7.7 EC-Standard P-256

ECDH is thus the key exchange used in Bluetooth [Blu]. In standard procedures based on ECDH, the parameters (p, r, s, G, o) are predefined and are thus effectively part of the algorithm. In order to show explicitly what such procedures look like, the NIST standard **P-256** used for Bluetooth is reproduced here [BeL, USG]:

- prime number p
  - p = FFFFFFFF 00000001 00000000 00000000 00000000 FFFFFFFF FFFFFFFF FFFFFFFF
- Elliptic curve $y^2 = x^3 + r \bullet x + s$ with
  - r = FFFFFFFF 00000001 00000000 00000000 00000000 FFFFFFFF FFFFFFFF FFFFFFFC = p − 3
  - s = 5AC635D8 AA3A93E7 B3EBBD55 769886BC 651D06B0 CC53B0F6 3BCE3C3E 27D2604B
- Base point $G = (x_G, y_G)$ of prime order o
  - x_G = 6B17D1F2 E12C4247 F8BCE6E5 63A440F2 77037D81 2DEB33A0 F4A13945 D898C296

Fig. 3.6 ElGamal cipher

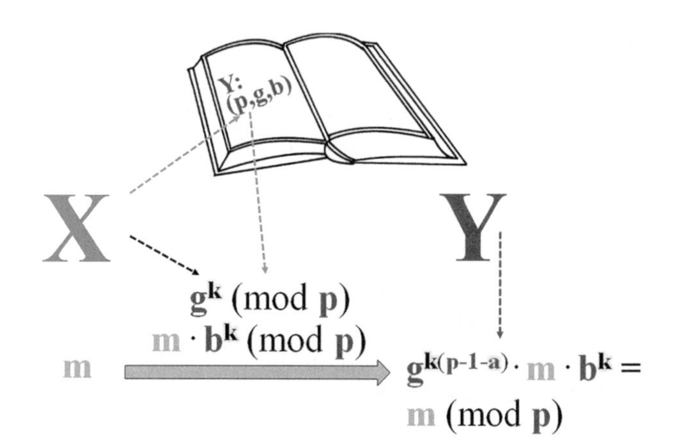

- y_G = 4FE342E2 FE1A7F9B 8EE7EB4A 7C0F9E16 2 BCE3357 6B315ECE CBB64068 37BF51F5
- o = FFFFFFFF 00000000 FFFFFFFF BCE6FAAD A7179E84 F3B9CAC2 FC632551

The parameters are represented hexadecimally, whereby 4 bits are combined into numbers from 0 to 15. The letters stand for the two-digit numbers A = 10, B = 11,…, F = 15.

3.7.8 Diffie-Hellman on the Internet

Even on the Internet, the case can arise that neither of the two communication partners has a valid RSA certificate for the key agreement. For this reason, the TLS handshake protocol also allows the AES key to be exchanged using the Diffie-Hellman method. Both the variant with discrete logarithm (DH) and the variant with elliptic curves (ECDH) are supported.

3.7.9 ECDH Security and Performance

In order to achieve a security level comparable to RSA or Diffie-Hellman, the key length of order o of the base point G is decisive for ECDH. The BSI guideline [BSI1] recommends at least 250 bits for this, as is also the case with P-256.

In general, the methods based on elliptic curves show a better runtime behavior even compared to RSA. In order to generate a value for o in the order of 250 bits, it is sufficient to use primes p with only about 250 bits, as can be seen in the example P-256. It is true that the addition of the points of an elliptic curve is relatively complex. But here one has to calculate only with remainders modulo prime numbers with about 250 bits and can therefore avoid the complex calculations with remainders in the order of 2000 bits.

3.8 ElGamal Cipher

How to transform the discrete logarithm problem into a public key cipher that can be used to encrypt arbitrary messages m is what we will now look at. The method originates from **Taher ElGamal** (born 1955) in 1984.

3.8.1 ElGamal Cipher

The **ElGamal cipher** is visualized in Fig. 3.6. Each potential receiver Y(ollanda) of messages registers in a central registry with her public key. To do this, she obtains a very large prime number p and a generating element g modulo p. She also chooses a natural number a in the range from 2 to p − 2 and computes $b = g^a \pmod{p}$. Finally, Y publishes as her **public key** (p, g, b) but keeps a secret as her **private key**. Thus, to obtain the private key a, an attacker A(rchibald) would have to solve the discrete logarithm of b to the base g modulo p.

Now suppose sender X(avier) wants to send a secret message m to receiver Y again. Then X looks up the public key (p, g, b) of Y in the central register and in turn chooses a random natural number k in the range from 2 to p − 2. Let the message m be a natural number smaller than the very large p. Sender X then transmits the encrypted information $g^k \pmod{p}$ and $m \cdot b^k \pmod{p}$ to Y. She in turn takes her private key a and computes $(g^k)^{(p-1-a)} \bullet m \bullet b^k = m \bullet (g^k)^{(p-1-a)} \bullet (g^a)^k = m \bullet g^{(p-1)k - ka + ak} = m \bullet g^{(p-1)k} = m \pmod{p}$. Since m is less than p, she has thus obtained the desired plaintext message.

The ElGamal procedure is very similar to the Diffie-Hellman key exchange, and the key selection even corresponds exactly to the semi-static Diffie-Hellman variant. For $g^k \pmod{p}$ it is the exchange value of X and for $b^k \pmod{p}$ it is the mutually agreed Diffie-Hellman key, which “masks” the secret message m in the ElGamal cipher.

- With ElGamal two cipher values have to be calculated and sent, with RSA only one.
- With ElGamal, two modular exponentiations are required for ciphering, with RSA only one.
- With ElGamal a new random number must be generated each time, with RSA none.

3.8.2 Example: ElGamal Cipher

As an example [Wil], receiver Y(ollanda) chooses the prime number p = 107. As can be calculated, g = 2 is a generating element modulo p = 107. Y also chooses a = 51, which she keeps secret as her private key, and calculates $b = g^a = 2^{51} = 80 \pmod{107}$. Therefore, she registers (p, g, b) = (107, 2, 80) as her public key.

Let us now assume that sender X(avier) wants to send Y the message m = 83. He chooses k = 17 as random number, calculates $g^k = 2^{17} = 104 \pmod{107}$ as well as $m \cdot b^k = 83 \cdot 80^{17} = 74 \pmod{107}$ and therefore sends 104 and 74. Receiver Y uses her private key a = 51 and calculates $(g^k)^{(p-1-a)} \bullet m \bullet b^k = 104^{55} \bullet 74 = 83 \pmod{107}$ and receives the plaintext message 83 from X this way.

3.8.3 ElGamal Cipher and Key Exchange

Again, it is necessary to note that the public-key cipher ElGamal requires much more computing time than symmetric ciphers. Therefore, like RSA, it is only used for **key exchange** for symmetric ciphers. This is also the reason why in practice the message m, i.e. the binary expanded key of a symmetric cipher, is smaller than p.

3.8.4 ElGamal Cipher for Elliptic Curves

Analogous to the Diffie-Hellman key exchange, the ElGamal cipher can also be applied to elliptic curves. Let p again be a prime number and $y^2 = x^3 + r \bullet x + s$ an elliptic curve modulo p. Moreover, let G be a base point on the curve with order o. Y(ollanda) chooses a natural number in the range from 2 to o − 1 and computes the point $B = b \bullet G$. The public key of Y is then (p, r, s, G, o, B), the private one b.

If X(avier) wants to send the message m to Y, i.e. usually the key for a symmetric cipher, he must first interpret m as a point M on the elliptic curve, preferably as the x-coordinate of $M = (m, y_0)$. From the curve equation $y_0^2 = m^3 + r \bullet m + s$, it should then be possible to compute y_0 as a square root modulo p. There are efficient calculation methods for this [Kob], but they can only lead to the goal if such a square root exists at all according to the Euler criterion (Sect. 3.3). However, half of the remainders modulo p are square roots, so X has a good chance that $M = (m, y_0)$ is a curve point. But if not, he just tries the point $M_1 = (m + 1, y_1)$ and $y_1^2 = (m + 1)^3 + r \bullet (m + 1) + s$ and continues with m + i until he succeeds [Kob].

To send m or M, X then chooses a random natural number k in the range from 2 to o − 1 and sends the points $k \bullet G$ and $M + k \bullet B$ to Y. The latter uses her private key b and computes $(M + k \bullet B) - b \bullet (k \bullet G) = M + k \bullet B - k \bullet B = M = (m, y_0)$, i.e., in particular, the desired message m.

3.8.5 Other Public-Key Ciphers

As attractive as the idea of developing public-key methods on the basis of difficult mathematical problems sounds, it is sobering to find that the pool is relatively small. The **Rabin cipher** by **Michael Rabin** (born 1931) is worth mentioning, but it is not used in practice

[BNS, Buc]. It is based on the fact that computing "square roots" modulo a large number $n = p \cdot q$ is as difficult as factoring n. In contrast, **Robert McEliece**'s (b. 1942) cipher using error-correcting **Goppa codes** [BNS, Man] requires an exorbitantly large, currently impractical key length. To complete this chapter we want to explain a totally different, easily understandable method, which is, however, of more historical interest.

3.9 Knapsack and Merkle-Hellman Cipher

3.9.1 Packed Knapsack

The so-called **knapsack** deals with the following problem: Given are natural numbers $v_1, \ldots, v_n$, where some of the v_i can be equal. Now bits b_i are to be calculated for a natural number v such that $v = b_1 \bullet v_1 + \ldots + b_n \bullet v_n$. One imagines a knapsack of volume v, which one wants to fill optimally with provisions each of volume v_i without leaving any part of the knapsack unused. Admittedly, this idea requires a very cunning backpacker. In any case, the fact is that solving this problem is hard, at least for sufficiently large v_i and n. There may be a unique solution to this, but there may also be many or none at all.

3.9.2 The Super-Knapsack

The difficulty of the problem changes dramatically, however, if we imagine a **super-knapsack** $v_1, \ldots, v_n$ ("superincreasing knapsack"), where each v_i is larger than the sum of all preceding v_j. Indeed, then the problem is super-simple to solve: Starting from the largest v_n, one goes backwards until one finds the first v_j that is less than or equal to the given v. Then you calculate $v - v_j$ and keep going backwards until you find the first v_k that is less than or equal to $v - v_j$. Then consider $v - v_j - v_k$ and continue this until you reach 0, which allows you to write v as the sum of these v_i. If you do not arrive at 0, you know that there is no solution.

Here is a concrete example of this: it is $v_1 = 2$, $v_2 = 3$, $v_3 = 7$, $v_4 = 15$, and $v_5 = 31$ a super-knapsack. We choose $v = 24$. The first of the v_j, which is less than or equal to v, is $v_4 = 15$. We therefore consider in the next step $v - v_4 = 24 - 15 = 9$. Now we can continue with $v_3 = 7$ and calculate $v - v_4 - v_3 = 2$. This ultimately shows $v = v_1 + v_3 + v_4$. So, expressed with the bits b_i, the solution is 10110.

3.9.3 Merkle-Hellman Cipher

Now it is again a matter of turning the Knapsack problem into a public-key procedure. The idea for the **Merkle-Hellman cipher** originated in 1978 from **Ralph Merkle** (born 1952) and **Martin Hellman** (born 1945).

The potential receiver Y(ollanda) chooses a super-knapsack $v_1,\ldots, v_n$. She also chooses a number m greater than $v_1 + \ldots + v_n$, and a number a in the range from 1 to m − 1 that is coprime with m. Then Y can also compute $b = a^{-1}$ (mod m) by using the extended Euclidean algorithm to determine the multiple sum $1 = b \bullet a + x \bullet m$. Finally, Y computes the residues $w_i = a \bullet v_i$ (mod m), making $w_1,\ldots, w_n$ no longer a super-knapsack. As her public key, Y publishes the knapsack $w_1,\ldots, w_n$, her private key is (b, m), from which the super-knapsack $v_1,\ldots, v_n$ can again be computed inversely.

A message block in Merkle-Hellman cipher consists of digital strings $b_1\ldots b_n$ of length n. Thus, if sender X(avier) wants to send a message to Y, he computes $w = b_1 \bullet w_1 + \ldots + b_n \bullet w_n$ and sends the number w. Since $w_1,\ldots, w_n$ is not a super-knapsack, a potential attacker A(rchibald) has a hard time computing the message $b_1\ldots b_n$ from w and the w_i.

For receiver Y, however, this is easy. Namely, she calculates $v = b \bullet w = b_1 \bullet b \bullet w_1 + \ldots + b_n \bullet b \bullet w_n = b_1 \bullet v_1 + \ldots + b_n \bullet v_n$ (mod m). But since by assumption m is greater than $v_1 + \ldots + v_n$, even $v = b_1 \bullet v_1 + \ldots + b_n \bullet v_n$. Now all Y has to do is solve the super-knapsack $v_1,\ldots, v_n$ for the number v to determine the message $b_1\ldots b_n$.

3.9.4 Example: Merkle-Hellman Cipher

Receiver Y(ollanda) uses the super-knapsack $v_1 = 2$, $v_2 = 3$, $v_3 = 7$, $v_4 = 15$, and $v_5 = 31$ from the above example [Kob], chooses m = 61 and a = 17, and determines $b = a^{-1} = 18$ (mod m = 61). Her private key is (b, m) = (18, 61), and her public key is respectively obtained from the remainder $w_i = 17 \bullet v_i$ (mod m = 61) to $w_1 = 34$, $w_2 = 51$, $w_3 = 58$, $w_4 = 11$, and $w_5 = 39$. Thus, if X(avier) wants to send the message 01101, he sends the number w = 51 + 58 + 39 = 148 instead. To decrypt, Y first computes the remainder $v = b \bullet w = 18 \bullet 148$ (mod m = 61), obtaining v = 41. Then, for v = 41, she solves the super-knapsack $v_1 = 2$, $v_2 = 3$, $v_3 = 7$, $v_4 = 15$, and $v_5 = 31$, and obtains 41 = 3 + 7 + 31, hence exactly the message 01101.

3.9.5 Security of the Merkle-Hellman Cipher

Even with multiple transformations with different modules m and factors a, the super-knapsack $v_1,\ldots, v_n$ is modified only slightly, namely too little. Therefore, **Adi Shamir** (b. 1952) already in 1982 found a method that exploits this shortcoming and cracks the Merke-Hellman cipher with only one transformation in reasonable time. Shortly thereafter, **Leonard Adleman** (b. 1945) elaborated that Shamir's method also works for Merkle-Hellman ciphers with multiple transformations. Knapsack-based methods are therefore no longer considered secure.

4 Digital Signature

4.1 Man-in-the-Middle Attack and Authentication

4.1.1 Passive and Active Attack

So far, we have always implicitly assumed that a potential attacker A(rchibald) plays only a passive role. He is intent on undermining the confidentiality between sender X(avier) and receiver Y(ollanda) by eavesdropping on and decrypting the secret communication in order to use the acquired knowledge for his own purposes immediately afterwards or even after a time delay. The scenario of passive eavesdropping is shown in Fig. 4.1.

However, attackers can also play an active role. This is called a **man-in-the-middle attack**. In this case, an attacker A(rchibald) inserts himself into a possibly two-way communication between X(avier) and Y(ollanda) and plays the role of Y to X and the role of X to Y. This scenario is visualized in Fig. 4.2. In general, active intervention requires that the attacker can also decrypt the messages. This is because he can then specifically change the communication to his liking.

4.1.2 Authentication of Messages

Receiver Y(ollanda) of a message should always be sure that she has received exactly the message that sender X(avier) has really sent. In addition, she should be able to convince herself beyond doubt of the origin of the message, i.e., that the message actually originated from the specified sender X. For example, if a man-in-the-middle attacker A(rchibald) succeeds in converting the originating message "Secret meeting tomorrow 10 a.m. at my place, Xavier" into

O. Manz, *Encrypt, Sign, Attack*, Mathematics Study Resources,
https://doi.org/10.1007/978-3-662-66015-7_4

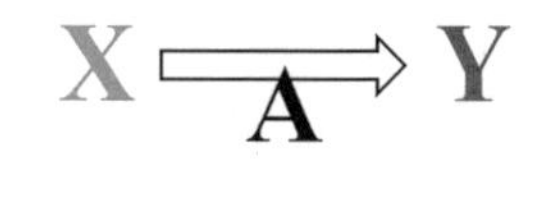

Fig. 4.1 Passive eavesdropping

Fig. 4.2 Man-in-the-middle attack

X A Y

- “Secret meeting tomorrow, ***11*** a.m. at my place, Xavier,” or
- “Secret meeting tomorrow, 10 a.m. at my place, ***Archibald.***”,

recipient Y should be able to detect these manipulations by taking appropriate precautions. This is called **authentication of messages** to protect against a man-in-the-middle attack.

4.1.3 MAC Authentication and CBC-MAC

A method for a fairly extensive authentication of a message is called **MAC** (Message Authentication Code). Such a MAC is often calculated using a symmetrical cipher E(•, •) in CBC mode (Sect. 2.6) with initialization value 0…0 and is then more precisely called **CBC-MAC**. Depending on a MAC key k_m the message (including sender identification data) is encrypted with CBC and only the last ciphertext block is appended to the message text m as checksum mac_{CBC} (m, k_m). The whole message is usually encrypted and sent with the same cipher E(•, •) but a different key k_c. The sender and receiver must therefore exchange both keys secretly, the key k_c for the cipher and the key k_m for the MAC. The receiver uses this to decrypt the received message, calculates the CBC-MAC for its part and authenticates by comparing it with the appended value.

A MAC procedure can therefore be used to secure messages against unauthorized changes. However, the check requires knowledge of the same secret key that was used to calculate the CBC-MAC. Therefore, anyone who can check a CBC-MAC is also able to calculate it. The recipient of a message can therefore verify for himself whether the supposed sender is really the originator and the message is authentic, but he cannot prove this to third parties. However, this is absolutely necessary for contractually binding documents and data records and can be achieved by means of a **digital signature** (Sect. 4.2).

WLAN (Sect. 3.2) and also Bluetooth (Sect. 3.7) use the so-called **AES-CCM method** (AES Counter Mode & CBC-MAC). With this method, encryption is performed with AES-128 in CTR mode and a CBC-MAC is appended to each WLAN or Bluetooth data packet, also using AES-128.

4.1.4 Authentication of Users

First of all, we would like to delimit the somewhat different concept of **authentication of users**. Here, a user usually registers with a central point of a system, the so-called **verifier**, which in turn authenticates the user on the basis of characteristics. Examples of this are the reading of a bank card with subsequent PIN entry at an ATM, the password entry when dialing into a computer network or the biometric passport check when entering and leaving the country at the airport. The procedures for authenticating a user generally rely on the following features, although combinations can also be used:

- **Knowledge** of a secret, e.g. PIN or password
- **Possession** of a personal document, e.g. identity card, company card, credit card
- **Biometrics**, e.g. fingerprint, voice, biometric passport photo

The focus of the authentication of a users is solely on the question of whether the user is currently really who he claims to be. In contrast to the authentication of messages, contents, for example the actions planned by a user, are of no importance. Furthermore, the authentication of a user only assesses his current situation, while old, archived messages also retain their authentication once it has been obtained.

Passive eavesdropping is sufficient to gain unauthorized knowledge of the features used for authentication; a man-in-the-middle attack is therefore not required. However, the verifier must know the respective characteristics of the features so that it can also check them. Thus, their secure and secret storage on the verifier's server is required. How to proceed with passwords, for example, and how to use the **digital signature** as an alternative will be discussed at the end of the chapter (Sect. 4.8).

Both message and user authentication are usually referred to briefly (and laxly) as **authentication**, and the meaning is only clear from context.

4.1.5 Attack on Diffie-Hellman Key Exchange

If X(avier) and Y(ollanda) agree on their key for a symmetric cipher using Diffie-Hellman key exchange, a man-in-the-middle attack is fatal. This is because attacker A(rchibald) can agree on a key with X in the role of Y and agree on a key with Y in the role of X without either of the communication partners X and Y being aware of it. Subsequently, A can tap messages from X, decrypt them, and forward them modified to Y, as well as vice versa. A Diffie-Hellman key exchange, such as with the TSL protocol on the Internet or with Bluetooth, is therefore exposed to a man-in-the-middle attack without any further precautions. In this case, it is therefore advisable for the two communication partners to authenticate themselves unambiguously to each other beforehand, i.e. not to a verifier. This is why, for example, the so-called **SSP procedure** (Secure Simple Pairing) has been implemented for Bluetooth, in which the users can first authenticate the devices communicating via Bluetooth by means of a six-digit number.

4.2 RSA and ElGamal Signature

In the classic paper-based processes, the legal authentication of a document is documented by the traceable signature of the responsible creator at the end of the text. To ensure that no pages can be exchanged in the case of a longer text, the creator also adds his or her initials to each page by hand. However, all this is no longer possible with digital message transmission.

4.2.1 Authentication by Digital Signature

We have dealt in detail with public-key ciphers in the last chapter, where participant Y(ollanda) registers her encryption key k_e publicly, since it is practically impossible to compute her private decryption key k_d from it. Thus, the key k_d is an absolute secret of Y. Let us denote by E(•, •) and D(•, •) the ciphering and deciphering, respectively, with an initially still arbitrary public-key cipher. In this case, if X(avier) wants to send a confidential message m to Y, X encrypts the message m into the ciphertext $c = E(m, k_e)$ using Y's public key k_e. Receiver Y decrypts using her private key k_d and receives the plaintext message $m = D(c, k_d)$.

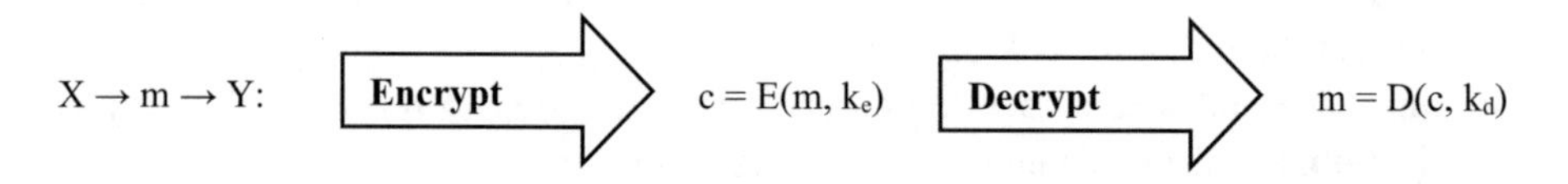

We now imagine as a scenario that in the context of a public key cipher two additional computation rules sig(•, •) and ver(•, •, •) would be specified, one for signing and one for verifying. Participant Y(ollanda) should be able to "sign" her message m in this scenario by computing a **digital signature** $s = sig(m, k_d)$ using her private key k_d. So only participant Y can generate this signature. Now Y sends the signature s together with the message m to participant X(avier), where we want to imagine m unencrypted for the moment. Participant X in this scenario should be able to verify the signature using Y's public key k_e by computing $ver(m, s, k_e)$. If the verification returns an "o. k.", then X accepts Y's signature as "valid".

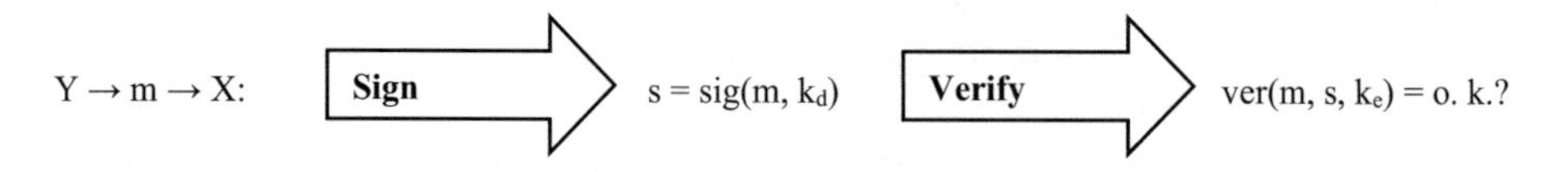

If the message text m had been changed by attacker A(rchibald) into another message m′, then ver(s, m′, k_e) would not result in an "o. k.". So we summarize the scenario again:

- Only participant Y(ollanda) can generate her signature s = sig(m, k_d), since only she knows her private key k_d.
- Participant X(avier) can use Y's public key k_e to verify the authenticity of the signature using ver(s, m, k_e).
- Once authenticity is authenticated, X can be sure that the message really came from Y and has not been altered along the way.
- X can even clearly prove this to any third party, such as Z(izi), since he could not have calculated the signature himself.

We will now look at how this scenario of a digital signature based on the public key methods RSA- and ElGamal can be realized.

4.2.2 RSA Signature

The **RSA signature** is quite simple and obvious, namely it uses exactly the same computational rules as the RSA cipher itself. Thus, participant Y(ollanda) obtains two different prime numbers p and q and multiplies them to n = p • q. She also chooses a natural number e less than (p − 1) • (q − 1), which is coprime with (p − 1) • (q − 1). Using the extended Euclidean algorithm, she determines a natural number d with 1 = d • e + b • (p − 1) • (q − 1). The public key of Y is then (n, e), her private is d. Let the message m again be a natural number smaller than n. Then $m^{ed} = m^{de} = m \pmod n$ (Sect. 3.1).

For the digital RSA signature of m, Y computes the remainder of m^d modulo n, which she sends to receiver X(avier) along with the plaintext m. The latter uses Y's public key (n, e) and the received value $m^d \pmod n$ and computes $m^{de} \pmod n$. Receiver X considers the signature verified if his computation produces exactly the received message $m = m^{de} \pmod n$. The procedure is shown schematically in Fig. 4.3.

Fig. 4.3 RSA signature

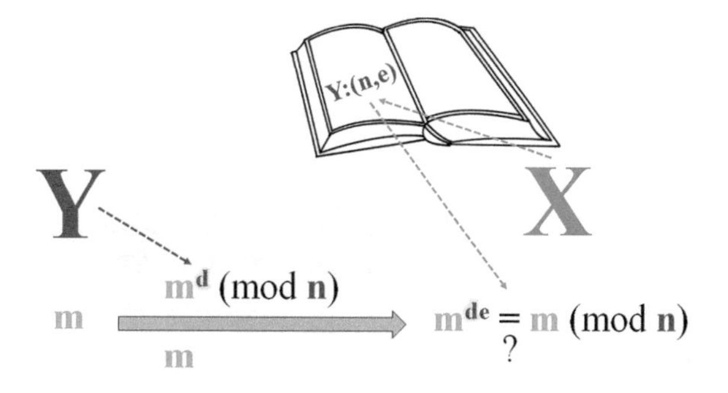

4.2.3 Example: RSA Signature

Here is an example [Hau1] for the RSA signature. Here Y(ollanda) chooses $n = p \cdot q = 13 \cdot 23 = 299$, and because of $(p - 1) \cdot (q - 1) = 264$ she can use $e = 5$ for her public key $(n, e) = (299, 5)$. Using the extended Euclidean algorithm, she computes $1 = 53 \cdot 5 - 1 \cdot 264$, so $d = 53$ is her private key.

If Y wants to sign the message $m = 296$, she calculates $m^d = 296^{53} = 212 \pmod{n = 299}$ (Sect. 3.1) and sends the signature 212 together with the message $m = 296$. Receiver X(avier) uses Y's public key (299, 5) and calculates $m^{de} = 212^5 = 296 \pmod{299}$. Since this results in $m = 296$, X accepts Y's RSA signature.

4.2.4 ElGamal Signature

The **ElGamal signature**, published by **Taher ElGamal** (b. 1955) together with his public key cipher in 1984, is a bit more complicated. Thus, participant Y(ollanda) obtains a prime number p and a generating element g modulo p. She also chooses a natural number a in the range from 2 to $p - 2$ and computes $b = g^a \pmod{p}$. Her public key is then (p, g, b), her private is a. Let the message m again be a natural number smaller than p.

For the digital ElGamal signature of m, Y chooses a random natural number k in the range from 2 to $p - 2$, which is coprime with $p - 1$, and also keeps this secret. Using the extended Euclidean algorithm, she computes a natural number x with $1 = x \cdot k \pmod{p - 1}$, so that $k^{-1} = x \pmod{p - 1}$ holds. As signature, along with the plaintext m, she sends the residues $u = g^k \pmod{p}$ and $v = (m - a \cdot u) \cdot k^{-1} \pmod{p - 1}$.

To verify the signature, receiver X(avier) computes $b^u \cdot u^v \pmod{p}$. He can do this because, on the one hand, he receives u and v and, on the other hand, he can look up b and p as Y's public key. Because of $b = g^a \pmod{p}$, $g^{p-1} = 1 \pmod{p}$, and $k^{-1} = x \pmod{p - 1}$,

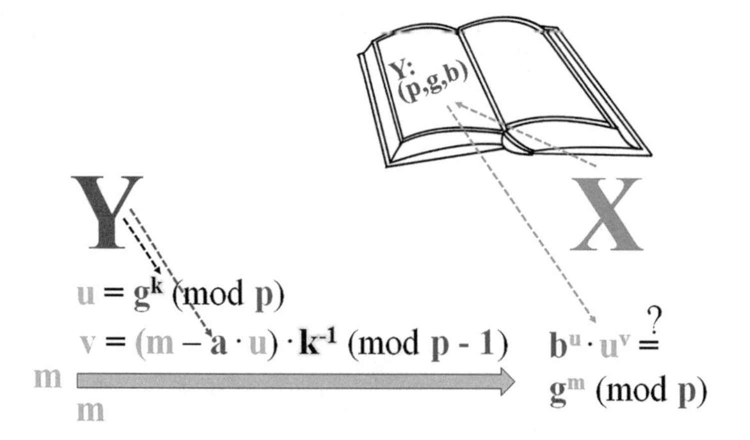

Fig. 4.4 ElGamal signature

he gets $b^u \bullet u^v = g^{au} \bullet g^{k(m-au)x} = g^{au} \bullet g^{m-au} = g^m \pmod{p}$. As a check, receiver X also computes the value $g^m \pmod{p}$ directly. He can do this since he receives m and can look up g as Y's public key. If both results are identical, X considers the signature verified. Figure 4.4 visualizes the procedure.

- With ElGamal two signature values have to be calculated and sent, with RSA only one.
- With ElGamal three modular exponentiations are necessary for verification, with RSA only one.
- With EGamal a new random number must be generated each time, with RSA none.

4.2.5 Example: ElGamal Signature

Here is also an example [Hau1] of the ElGamal signature. As one computes, $g = 3$ is a generating element modulo $p = 17$. Moreover, $3^{11} = 7 \pmod{17}$. Therefore, Y(ollanda) chooses as her ElGamal public key $(p, g, b) = (17, 3, 7)$ and as its private key $a = 11$.

If Y wants to sign the message $m = 10$, she chooses the random number $k = 13$ for this purpose, where 13 is coprime with $p - 1 = 16$. Using the extended Euclidean algorithm, she determines the multiple sum $1 = 5 \bullet 13 - 4 \bullet 16 = 5 \bullet k - 4 \bullet (p - 1)$, so $k^{-1} = 5 \pmod{p - 1 = 16}$. Finally, she computes the signature values $u = g^k = 3^{13} = 12 \pmod{p = 17}$ and $v = (m - a \bullet u) \bullet k^{-1} = (10 - 11 \bullet 12) \bullet 5 = 14 \pmod{p - 1 = 16}$ and sends the signature values $u = 12$ and $v = 14$ to X(avier) together with the message $m = 10$.

For verification, X computes both $g^m = 3^{10} = 8 \pmod{p = 17}$ and $b^u \bullet u^v = 7^{12} \bullet 12^{14} = 13 \bullet 15 = 8 \pmod{p = 17}$. Since both values match, X accepts Y's ElGamal signature.

4.2.6 Cryptographic Envelope and Digital Fingerprint

The signature of a message m is usually abbreviated to **sig(m)**, irrespective of the actual procedure used. In our previous signature scheme, the plaintext m was sent unencrypted and thus unprotected. Of course, we still have to change this. As we know, to encrypt messages, one always uses a symmetric cipher because of its better performance, which we will briefly denote here by $S(\bullet, \bullet)$ and its key by k. So in principle one could encrypt m together with its signature sig(m) using $S(\bullet, \bullet)$ and therefore send $S(m||sig(m), k)$, the so-called **cryptographic envelope**.

The signature sig(m), however, is based on a public-key cipher, which we already know to be much too slow for practical applications when applied to a complete message m (more precisely, to its binary expansion block by block). We have therefore used public-key ciphers only for key exchange for symmetric ciphers. As good as the idea of digital signatures is, is this already its very end? Not quite: We need a method to derive a **digital fingerprint** from any message, which on the one hand is only a few bits long, i.e. of the order of magnitude of the keys of symmetric ciphers, but on the other hand uniquely characterizes the message. This fingerprint can then be signed instead of the complete message m in the cryptographic envelope.

4.3 Hash Value and Secure Hash Algorithm SHA

4.3.1 Cryptographic Hash Function and Hash Value

We now imagine a message m again as a bit sequence of in principle arbitrary length. We call h(•) a **hash function** of **length** n, if it maps bit strings of arbitrary finite length to bit strings of a fixed length n. For reasons of practicability, it is required that this computation can be performed efficiently, i.e., the hash value h(m) should be able to be generated for any arbitrarily long message m in a reasonable amount of time (so-called **computability**). One speaks more precisely of a **cryptographic hash function** if the following two properties are also fulfilled:

- It should not be possible to find two different messages m and m′ with the same hash value h(m) = h(m′) in a reasonable amount of time (so-called **collision resistance**).
- It should not be possible to find a message m with hash value h(m) = y for a randomly chosen bit string y of length n in a reasonable amount of time (so-called **one-way property**).

Digital fingerprints are thus generated by means of cryptographic hash functions whose length should be at most a few hundred bits. Then, on the one hand, it follows from the computability that the **hash value** h(m) and thus also the signature sig(h(m)) of the binary expanded hash value can be computed much faster than the signature sig(m) of the entire message m. On the other hand, the collision resistance ensures that the hash value h(m) uniquely characterizes the message m in principle and thus makes a meaningful signature possible in the first place. The one-way property primarily concerns the RSA signature. In this case, an attacker can determine a y for a given z via verification with sig(y) = z. If he were also additionally able to efficiently compute a message m with h(m) = y, he might falsely claim that z = sig(h(m)) would be a valid signature for m. Where the one-way property is still needed, we see in the secure storage of passwords as their hash values (Sect. 4.8).

4.3.2 MAC Authentication and HMAC

We will mention another application of hash functions in a moment. They can also be used as a MAC (Sect. 4.1) for the authentication of a message m and are then referred to as an **HMAC**. It would be obvious to use a hash function h(•) and a MAC key k_m to calculate the hash value $h(k_m||m)$. However, this procedure is considered to be insecure.

Therefore, a somewhat more complicated procedure is used as HMAC. Here k_m must have the bit length n of the hash function h(•) and n must be divisible by 8. Furthermore, let the two constants c_1 and c_2 be defined hexadecimally by $c_1 = 5C\ldots_{n/8}\ldots 5C$ and $c_2 = 36\ldots_{n/8}\ldots 36$. Then the HMAC of m is calculated as $mac_H(m, k_m) = h((k_m \oplus c_1)||h((k_m \oplus c_2)||m))$ [WPHMA].

The HMAC is appended to the message text m as a checksum in the same way as the CBC-MAC, the whole thing is symmetrically encrypted with a key k_c and sent. The recipient decrypts, verifies and authenticates the message with k_c and k_m.

4.3.3 Merkle-Damgård Construction

Let us first clarify the principle of one of the most common hash function construction methods, the **Merkle-Damgård construction**, which dates back to work by **Ralph Merkle** (b. 1952) and **Ivan Damgård** (b. 1956). One iteratively constructs a hash function h(•) of length n using the Merkle-Damgård construction, but requires a **compression function** F(•) that maps bit sequences of length n + r to bit sequences of length n for a suitable natural number r. We will see below how F(•) can be chosen. The hash value h(m) of any message m is then computed in the Merkle-Damgård construction using the compression function F(•) as follows: First, one decomposes the message $m = m_1\ldots m_t$ into t bit blocks m_i of length r, appropriately padding at the end for each concrete procedure. Then one starts with an initial bit sequence h_0 of length n, which is also specifically determined for each concrete procedure, and calculates the value $h_1 = F(h_0||m_1)$ for the bit sequence $h_0||m_1$ of length n + r. The compression function F(•) maps the bit sequence $h_0||m_1$ of length n + r into a bit sequence h_1 of length n. In the next step, one computes $h_2 = F(h_1||m_2)$ and in general $h_i = F(h_{i-1}||m_i)$. Finally, the last result h_t is the hash value h(m) of length n for message m. The diagram in Fig. 4.5 illustrates the Merkle-Damgård construction again [WPMDK].

The 1979 construction method was originally proposed by Ralph Merkle. Ivan Damgård proved in 1989 that if the message m is suitably prepared, a collision-resistant compression function F(•) leads to a collision-resistant hash function h(•).

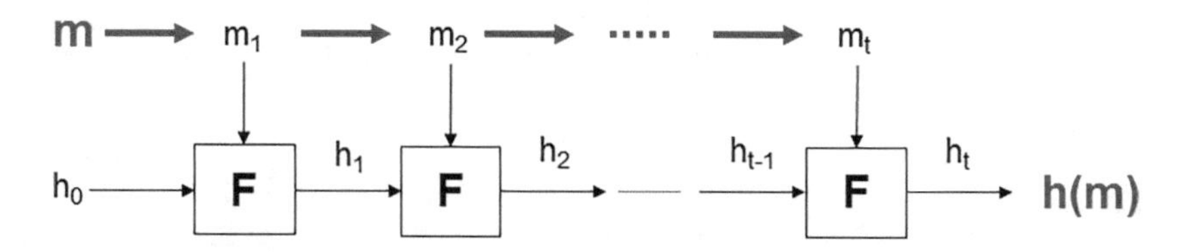

Fig. 4.5 Merkle-Damgård construction of a hash function

4.3.4 Davies-Meyer and Matyas-Meyer-Oseas Compression Function

Thus, to concretely specify a hash function of length n using the Merkle-Damgård construction, it suffices to specify the compression function F(•). The main component of F(•) is usually a block cipher E(•, •). This can be a standard cipher such as Triple-DES or AES, but an individually constructed cipher is usually used. Indeed, in hash functions, one places particular emphasis on simple, fast operations that can be implemented efficiently. We will look at an example, namely SHA-2-256 (Sect. 4.3 end).

The **Davies-Meyer compression function** F_{DM} (•) uses the message block m_i of bit length r as the key for the cipher E(•, •), and the preceding iterated hash value h_{i-1} of length n as its plaintext. The ciphertext is then still added bitwise ⊕ to h_{i-1}, to compute the next iterated hash value h_i. The compression function is thus $F_{DM}(h_{i-1}||m_i) = E(h_{i-1}, m_i) \oplus h_{i-1}$, which is visualized schematically in Fig. 4.6.

Alternatively, in the Davies-Meyer compression function, one also decomposes $h_{i-1} = h_{i-1}^{(1)}\ldots h_{i-1}^{(n/w)}$ and $E(h_{i-1}, m_i) = c_i^{(1)}\ldots c_i^{(n/w)}$ into components $h_{i-1}^{(j)}$ and $c_i^{(j)}$ of smaller bit length w and computes $h_i = h_i^{(1)}\ldots h_i^{(n/w)}$ component-wise as $h_i^{(j)} = c_i^{(j)} \boxplus h_{i-1}^{(j)}$ by adding ⊞ modulo 2^w, where the bit strings of length w are interpreted as a binary expansion of a natural number.

The **Matyas-Meyer-Oseas compression function** F_{MMO} (•) proceeds approximately in reverse. However, the prerequisite for this is that the bit length r of the message blocks m_i is chosen to be equal to the length n of the hash value. The message block m_i is then used as the plaintext block for the block cipher E(•, •), and the ciphertext is subsequently added to m_i bitwise ⊕, so as to compute the next iterated hash value h_i. The previous iterated hash value h_{i-1} is used as the key for the block cipher. However, this only works if the block cipher has equal block and key length. If this is not the case, h_{i-1} is first made suitable using a suitable function G(•). Thus, the Matyas-Meyer-Oseas compression function is $F_{MMO}(h_{i-1}||m_i) = E(m_i, G(h_{i-1})) \oplus m_i$. Figure 4.7 again visualizes this schematically.

4.3.5 Cryptographic Hash Functions SHA

The cryptographic hash functions most frequently used in practice today are those of the so-called **SHA family** (Secure Hash Algorithm). The first generation **SHA-1** has a length of 160 bits and was standardized by NIST in 1995. It is based on a Merkle-Damgård construction together with a Davies-Meyer compression function. However, one did not use

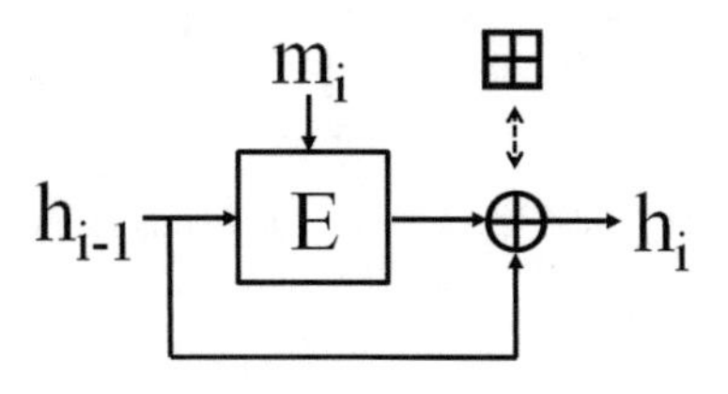

Fig. 4.6 Davies-Meyer compression function

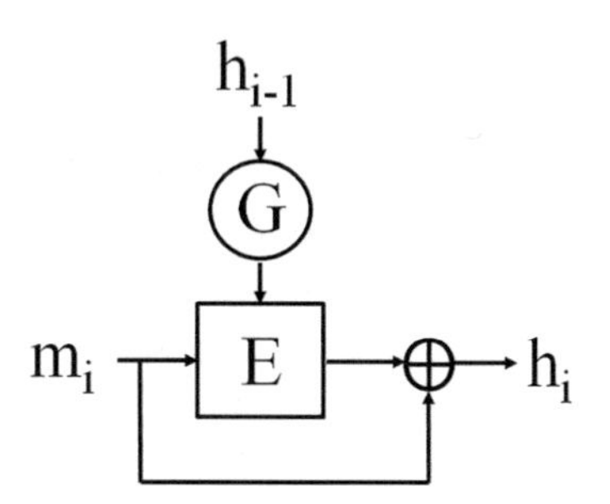

Fig. 4.7 Matyas-Meyer-Oseas compression function

an existing, possibly standardized block cipher, but developed it individually for SHA-1. By 2004, there had been several successful attacks against SHA-1, and it was discovered that SHA-1 is far less collision resistant than had been theoretically expected.

In response to the attacks that became known, NIST held a workshop in 2005 to discuss the current status of hash functions. NIST recommends the transition to **SHA-2** hash functions of the second generation. These are the variants SHA-2-224, SHA-2-256, SHA-2-384 and SHA-2-512, where the appended number indicates the length of the hash value in each case. It is also based on a Merkle-Damgård construction together with a Davies-Meyer compression function, although the block cipher has been modified in SHA-2 compared to SHA-1 [WPSH2]. There have been no relevant attacks on SHA-2 so far, so SHA-2 may still be considered secure with the exception of the smallest variant SHA-2-224. However, if SHA-2 should also turn out to be compromised or insecure, there was initially no other standardized hash function available that was recognized as secure.

Therefore, it was decided to create a new standard that would take into account current research. In order to standardize a hash function with a different construction principle, NIST organized a tender in 2007 along the lines of the AES. The choice was made in 2012 for the method called **Keccak**, which was standardized in 2015 as **SHA-3** with variants SHA-3-224, SHA-3-256, SHA-3-384, and SHA-3-512. SHA-3 is constructed in a fundamentally different way than SHA-2, namely with the help of a so-called **sponge construction** [WPSH3].

4.3.6 Cryptographic Hash Functions MD

MD5 (Message-Digest-Algorithm 5) is also a widely used cryptographic hash function, but it only produces a 128-bit hash value. MD5 was developed in 1991 by **Ronald Rivest** (born 1947) after it became clear that the previous version **MD4** was insecure. However, MD5 is not considered very secure today either. A successor version **MD6** with hash length 256 bits was submitted by Rivest to the NIST tender in 2009, but did not reach the second round of the proceedings and therefore does not play a significant role today.

4.3.7 Cryptographic Hash Function SHA-2-256

Since the SHA-2 hash functions have now become the de facto standard, let us take a closer look at **SHA-2-256** [IWS, WPSH2e]. It is a 256-bit hash value computed using a Merkle-Damgård construction together with the second variant of a Davies-Meyer compression function. The message $m = m_1 \ldots m_t$ is thereby split into blocks m_i of length 512 bits, with padding at the end according to rules not described in more detail here. Thus, we must first describe how the block cipher $E(h_{i-1}, m_i)$ is constructed for an arbitrary index i and the iterated hash value h_{i-1}. To do this, consider the 256 bits of h_{i-1} composed as $h_{i-1} = h_{i-1}^{(1)} \| \ldots \| h_{i-1}^{(8)}$ with eight blocks $h_{i-1}^{(1)}, \ldots, h_{i-1}^{(8)}$ of 32 bits each. Namely, the cipher $E(\bullet, \bullet)$ operates on eight 32-bit blocks, has 64 rounds, and uses only simple and fast operations:

$\oplus$	Bitwise addition
$\circ$	Bitwise multiplication
$\neg$	Bitwise NOT (i.e. $\neg 0 = 1, \neg 1 = 0$)
R_r^k	Cyclic shifting of a bit string to the right by k positions
S_r^k	Shifting a bit string to the right by k positions, padded with 0
$\boxplus$	Addition modulo 2^{32} of 32-bit strings (interpreted as natural number)

Derived from this, the following operations are required for bit strings x, y, z of length 32:

$$C(x,y,z) = (x \circ y) \oplus (\neg x \circ z)$$
$$M(x,y,z) = (x \circ y) \oplus (x \circ z) \oplus (y \circ z)$$
$$\Sigma_0(x) = R_r^2(x) \oplus R_r^{13}(x) \oplus R_r^{22}(x)$$
$$\Sigma_1(x) = R_r^6(x) \oplus R_r^{11}(x) \oplus R_r^{25}(x)$$
$$\sigma_0(x) = R_r^7(x) \oplus R_r^{18}(x) \oplus S_r^3(x)$$
$$\sigma_1(x) = R_r^{17}(x) \oplus R_r^{19}(x) \oplus S_r^{10}(x)$$

Figure 4.8 shows how, for the total of 64 rounds of $E(\bullet, \bullet)$, the j. round for $j = 0, \ldots, 63$ is constructed. Here, a,…, h denote the placeholders of 32 bits each, which are initialized with $a = h_{i-1}^{(1)}, \ldots, h = h_{i-1}^{(8)}$ and are recalculated and filled accordingly for each round.

The w_j result from the message block m_i. Namely, it is $m_i = w_0 w_1 \ldots w_{15}$ the decomposition of the message block m_i into 16 sub-blocks w_j with 32 bits each. The remaining w_j for $j = 16, \ldots, 63$ are recursively computed from $w_j = \sigma_1(w_{j-2}) \boxplus w_{j-7} \boxplus \sigma_0(w_{j-15}) \boxplus w_{j-16}$. The constants $k_0, \ldots, k_{63}$, are as follows:

```
428a2f98 71374491 b5c0fbcf e9b5dba5 3956c25b 59f111f1
923f82a4 ab1c5ed5 d807aa98 12835b01 243185be 550c7dc3
```

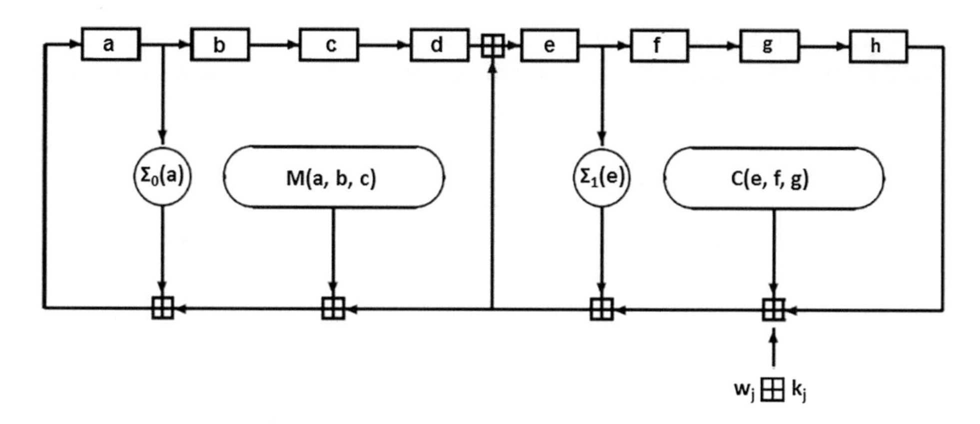

Fig. 4.8 Round function of SHA-2-256

```
72be5d74 80deb1fe 9bdc06a7 c19bf174 e49b69c1 efbe4786
0fc19dc6 240ca1cc 2de92c6f 4a7484aa 5cb0a9dc 76f988da
983e5152 a831c66d b00327c8 bf597fc7 c6e00bf3 d5a79147
06ca6351 14292967 27b70a85 2e1b2138 4d2c6dfc 53380d13
650a7354 766a0abb 81c2c92e 92722c85 a2bfe8a1 a81a664b
c24b8b70 c76c51a3 d192e819 d6990624 f40e3585 106aa070
19a4c116 1e376c08 2748774c 34b0bcb5 391c0cb3 4ed8aa4a
5b9cca4f 682e6ff3 748f82ee 78a5636f 84c87814 8cc70208
90befffa a4506ceb bef9a3f7 c67178f2
```

The k_j are represented hexadecimally, whereby 4 bits each are combined to form numbers from 0 to 15. The letters stand for the two-digit numbers a = 10, b = 11,…, f = 15.

As a result of the 24 rounds, the procedure thus yields eight blocks $c_i^{(1)}, \ldots, c_i^{(8)}$ with 32 bits each, from which $E(h_{i-1}, m_i) = c_i^{(1)}\|\ldots\|c_i^{(8)}$ is composed. According to the second variant of the Davies-Meyer compression function, h_i is finally calculated as $h_i = h_i^{(1)}\|\ldots\|h_i^{(8)}$, where $h_i^{(1)} = h_{i-1}^{(1)} \boxplus c_i^{(1)}, \ldots, h_i^{(8)} = h_{i-1}^{(8)} \boxplus c_i^{(8)}$ applies. The initial hash value $h_0 = h_0^{(1)}\|\ldots\|h_0^{(8)}$ is

```
    6a09e667 bb67ae85 3c6ef372 a54ff53a 510e527f 9b05688c 1f83d9ab
5be0cd19
```

4.4 Email with PGP and WhatsApp

4.4.1 Sending and Receiving Cryptographic Envelopes

It remains to be clarified and described what is really calculated and sent in messages in a cryptographic envelope. You only sign the hash value, i.e. the digital fingerprint of a message. And this is how the entire procedure looks in principle when sending and receiving:

- Sender Y(ollanda) chooses a symmetric cipher S(•, •), a public-key cipher E(•, •), a digital signature sig(•), and a hash function h(•), which she agrees on in advance with receiver X(avier).
- Moreover, Y generates the cipher key k of the symmetric cipher S(•, •) using a random generator, looks up the public key e of X for the public key cipher E(•, •), and computes E(k, e).
- Now Y calculates the hash value h(m) for message m and signs it with her private signature key.
- Finally, to send m secret and signed to receiver X, Y transmits the cryptographic envelope S(m||sig(h(m)), k) together with E(k, e).
- X first deciphers E(k, e) with his private key and then S(m||sig(h(m)), k) with key k. He thus receives the message m together with the signature of the hash value h(m).
- Now he in turn computes the hash value h(m) of the received message m.
- Finally, X looks up Y's public signature key and uses it to verify the received signature sig(h(m)), i.e. he checks whether the verification function for h(m), sig(h(m)) and the public key does yield an o.k.
- If this is successful, receiver X considers both sender Y and message m to be authenticated.

4.4.2 PGP Pretty Good Privacy

We want to concretize the procedure using the example of e-mails. The **PGP** (Pretty Good Privacy) program package is used to encrypt and authenticate data and is primarily used for e-mails. It was originally written by **Phil Zimmermann** (born 1954) and first published in 1991. PGP is a program package that uses both symmetric and public-key ciphers. The following description of how an e-mail is encrypted and digitally signed with PGP is schematically visualized in Fig. 4.9.

- To ensure that the message cannot be tampered with and to unambiguously prove the sender, PGP generates a digital signature of the entire e-mail m. First, the hash value h(m) is calculated using a hash function h(•). This creates a unique digital fingerprint

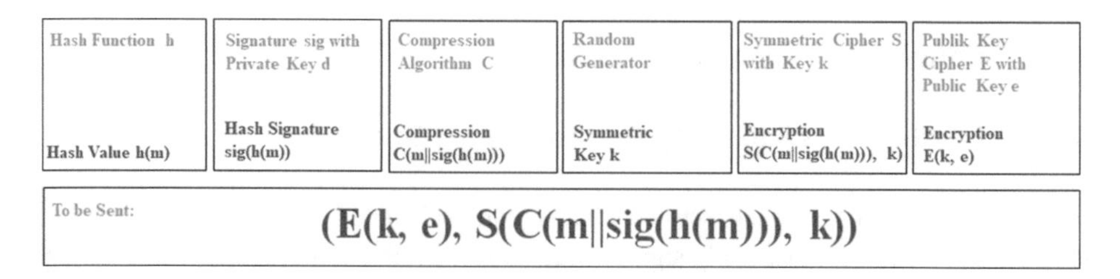

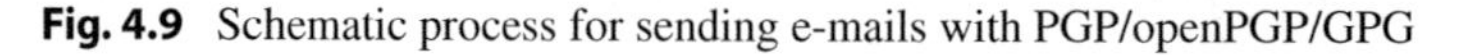

Fig. 4.9 Schematic process for sending e-mails with PGP/openPGP/GPG

that is much shorter than m itself. Subsequently, the sender's digital signature sig(h(m)) is generated for this hash value using the sender's private signature key.
- Now PGP can perform the encryption of the e-mail. To do this, m is first combined with the digital signature sig(h(m)) to form a data record and subjected to data compression C(•). On the one hand, this reduces the size of the data set and, on the other hand, makes cryptanalysis more difficult by reducing, for example, linguistic redundancy. The compressed data C(m||sig(h(m))) are now encrypted with a symmetric cipher S(•, •) and a randomly generated key k to form a ciphertext S(C(m||sig(h(m))), k).
- The randomly generated key k must also be communicated to the receiver. To do this, k is encrypted using a public-key cipher E(•, •) and the recipient's public key e to form E(k, e) and prefixed to the symmetrically encrypted ciphertext S(C(m||sig(h(m))), k). This entire packet is finally sent under PGP.

4.4.3 OpenPGP and GPG

The PGP program was sold several times over the course of time to various software houses, from which licenses could be purchased. In 2010, the software was transferred to the US company Symantec. However, Phil Zimmermann already published the complete PGP source code in 1995 in the book "PGP Source Code and Internals". This was painstakingly typed out, and on the basis of this the freely available **OpenPGP** standard was developed and maintained in parallel with open source software packages. Initiated by **Werner Koch** (born 1961), the first version of **GPG** (GNU Privacy Guard) was published in 1997. This is also a freely available system developed on the basis of OpenPGP, comparable to PGP in structure and range of functions and largely compatible [WPGPG, WPOPG].

4.4.4 Implemented Procedures for PGP/openPGP/GPG

Originally, PGP used DES as the symmetric cipher. In the meantime, however, PGP/openPGP/GPG offers, among others, Triple-DES and AES with 128- or 256-bit keys, each of which is operated in CFB mode and therefore as a stream cipher with a pseudo-random sequence. The public key cipher for the exchange of the cipher key and for the digital signature was only RSA in the original PGP version. In the meantime, PGP/openPGP/GPG can also use ElGamal and ECDH with the P-256 standard for key exchange. SHA-2-256 and MD5, among others, can be used as hash functions, and data compression is performed using the zip-format.

For digital signatures, PGP/openPGP/GPG now also provides the DSA and ECDSA methods as well as the important secp256k1, brainpoolP256r1 and Curve25519 standards for elliptic curves (which will be explained in Sect. 4.5).

4.4.5 WhatsApp

But first we want to talk about another popular communication service besides SMS and email, namely **WhatsApp**. WhatsApp was founded in 2009 and has been part of Facebook since 2014. Users can exchange text messages as well as image, video and sound files between two people or in groups via WhatsApp.

WhatsApp has had a comprehensive security concept [WhA] since 2016. This is basically designed as follows: First, an ECDH key, the so-called identity key, is generated for each subscriber during the WhatsApp installation for a semi-static Diffie-Hellman key exchange based on elliptic curves (Sect. 3.7). The public part of the identity key is transferred to the WhatsApp server, the private part cannot be accessed. For ECDH, the elliptic curve according to Standard Curve25519 is used (Sect. 4.5).

In order to establish a protected WhatsApp communication, the sender performs a semi-static ECDH key exchange with the recipient. Both thus have a shared secret, the so-called master secret, from which a so-called root key is derived. When a WhatsApp message is to be sent, another ECDH key exchange is performed between the sender and the recipient, and a so-called chain key is derived from this using the root key, from which a so-called message key is formed using a hash function. The first substring of 256 bits of this message key is used as the key for an AES cipher in CBC mode, which is used to encrypt the WhatsApp message. Unlike PGP, WhatsApp does not use a digital signature to authenticate messages, but rather an HMAC (Sect. 4.3) based on the SHA-2-256 hash function and a 256-bit MAC key derived from the next substring of the message key. In contrast to PGP, which can be used in particular to send contractual documents, the authenticity of a message cannot be proven to third parties with WhatsApp.

Because of the vulnerability of Diffie-Hellman and thus also ECDH against a man-in-the-middle attack, a procedure for authentication of the communication partners has also been implemented for WhatsApp, similar to Bluetooth. A hash value is generated from the participant name and identity key using SHA-2-512, which can be scanned in as a QR code for verification.

Facebook is also planning a similar security concept for its **Instagram** photo and video sharing social networking service.

4.5 DSA and ECDSA Signature

4.5.1 Discrete Logarithm with Non-Generating Element

Let again p be a prime number and g a generating element modulo p. If q is a prime number dividing $p - 1$, then we can compute $h = g^{(p-1)/q} \pmod p$. Then $h^q = 1 \pmod p$, and the powers $1 = h^0, h^1, h^2, h^3, \ldots, h^{q-1}$ run through exactly q distinct values modulo p. To given b, one again calls the exponent a in $b = h^a \pmod p$ the **discrete logarithm** of b, but this time to the base of the nongenerating element h modulo p. It is believed that for sufficiently large q this problem is as difficult as the discrete logarithm to the base g, although in fact the h^i (mod p) run through less distinct elements than the g^j (mod p), namely q instead of p − 1. In any case, it is a fact that all known methods for calculating discrete logarithms also have an unrealistic large running time with respect to the base h as they have with respect to the base g, provided, however, that q has not been chosen too small.

4.5.2 DSA Signature

In practice, the ElGamal signature (Sect. 4.2), which is more important from a historical point of view, is hardly ever used. Instead, the standard procedure **DSA** (Digital Signature Algorithm) is usually used. This is a more efficient variant of the ElGamal signature, which was first standardised by NIST in 1994, with the latest revision dating from 2013.

Participant Y(ollanda) chooses prime numbers p and q such that q is a divisor of p − 1. For a generating element g modulo p, she computes $h = g^{(p-1)/q}$ (mod p), chooses a natural number a in the range from 2 to q − 1, and computes $b = h^a$ (mod p). Her public key is then (p, q, h, b), her private key is a.

For the DSA signature of a message m, Y chooses a random natural number k in the range from 2 to q − 1, which is thus automatically coprime with the prime number q, and also keeps this secret. Using the extended Euclidean algorithm, she computes a natural number x with $1 = x \bullet k$ (mod q), so that $k^{-1} = x$ (mod q) holds. As her signature, together with the plaintext m, she sends on the one hand the residue $u = (h^k \text{ (mod p)}) \text{ (mod q)}$, where here first modulo p and then modulo q is calculated, as well as the remainder $v = k^{-1} \bullet (m + a \bullet u)$ (mod q). If u = 0 or v = 0 holds, Y starts again with another random number k.

Receiver X(avier) verifies the authenticity of the signature values u and v as follows: Using the extended Euclidean algorithm, he computes a natural number y for v with $1 = y \bullet v$ (mod q), so $v^{-1} = y$ (mod q) holds. This is possible because the prime number q is part of the public key of Y and v is not 0, and therefore is coprime with q. Using the received message m, he computes the residues $w = m \bullet v^{-1}$ (mod q) and $w' = u \cdot v^{-1}$(mod q) and uses Y's public key (p, q, h, b) to check whether the residue $\left(h^w \cdot b^{w'} \left(\bmod p\right)\right)\left(\bmod q\right)$ equals the first signature value u. If it is, X considers the signature to be verified. This is because if the DSA signature is correct, then due to $h^q = 1$ (mod p), it follows that $h^w \cdot b^{w'} = h^{my} \cdot b^{uy} = h^{y(m+au)} = h^{y(vk)} = h^k \left(\bmod p\right)$ holds. In particular, u is then equal to $\left(h^w \cdot b^{w'} \left(\bmod p\right)\right)\left(\bmod q\right)$. Figure 4.10 visualizes the procedure.

The NIST standard specifies values for p and q in the order of 1024 and 160 bits, 2048 and 224 bits, 2048 and 256 bits, and 3072 and 256 bits, respectively, although 1024 and 160 bits are no longer recommended today.

In our description, we have signed the message m itself for the sake of simplicity. However, as we know, you actually sign a hash value of m. In the original NIST standard, SHA-1 was intended for this purpose, but this is no longer considered completely secure. Therefore, SHA-2 has also been approved in the meantime.

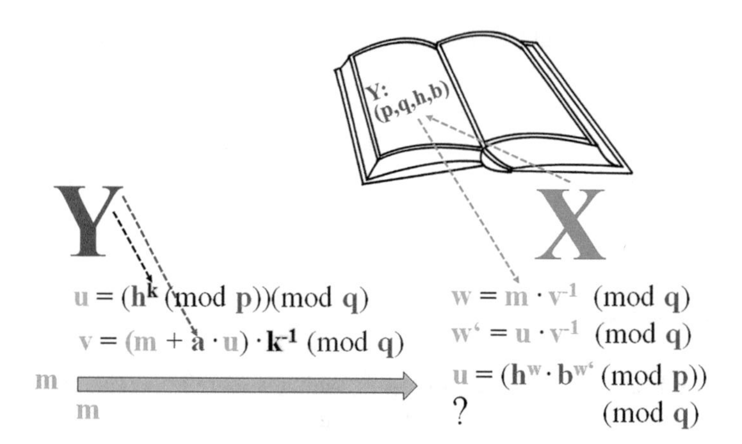

Fig. 4.10 DSA signature

Although DSA at first seems more complicated than the ElGamal signature, it requires only two instead of three modular exponentiations for verification. Moreover, the exponents as well as the signature values to be sent are also significantly smaller, since they are only remainders modulo the smaller prime q.

4.5.3 Example: DSA Signature

As an example [DIM], participant Y(ollanda) chooses prime numbers $p = 283$ and $q = 47$, where q divides $p - 1 = 282$. Also, Y chooses the number $h = 60$, for which $h^q = 60^{47} = 1$ (mod $p = 283$), and $a = 24$, and computes $b = h^a = 60^{24} = 158$ (mod $p = 283$). Therefore, she registers $(p, q, h, b) = (283, 47, 60, 158)$ as her public key, and her private key is $a = 24$.

Now suppose Y wants to send the message $m = 41$ to receiver X(avier) and also wants to sign it digitally. Then Y chooses a random number $k = 15$ and calculates $x = k^{-1} = 15^{-1} = 22$ (mod $q = 47$). Now Y first computes $h^k = 60^{15} = 207$ (mod $p = 283$) and then the remainder $u = 207 = 19$ (mod $q = 47$). Finally, since u is not equal to 0, Y also computes the remainder $v = (m + a \bullet u) \bullet k^{-1} = (41 + 24 \bullet 19) \bullet 22 = 30$ (mod $q = 47$). Since v is also not equal to 0, Y sends the signature values $u = 19$ and $v = 30$ to X along with the plaintext $m = 41$.

Receiver X first calculates $y = v^{-1} = 30^{-1} = 11$ (mod $q = 47$). With this in turn, he calculates the residues $w = m \cdot v^{-1} = 41 \cdot 11 = 28$(mod $q = 47$) und $w' = u \cdot v^{-1} = 19 \cdot 11 = 21$(mod $q = 47$). This ultimately shows that $h^w \cdot b^{w'} = 60^{28} \cdot 158^{21} = 106 \cdot 42 = 207(\bmod p = 283)$ and hence $207 = 19$ (mod $q = 47$). Since this yields the signature value $u = 19$ as a remainder, X considers Y's DSA signature to be verified.

4.5.4 Diffie-Hellman with Non-Generating Element

At this point a reference to the Diffie-Hellman key exchange is necessary (Sect. 3.5). In practice this can and is usually done with a non-generating element $h = g^{(p-1)/q} \pmod p$ instead of the generating element g modulo p. So the prime numbers p and q as well as h are publicly known. The procedure then looks quite analogously as follows:

- X(avier) chooses a random natural number e in the range from 2 to $q - 1$.
- Y(ollanda) chooses a random natural number f in the range from 2 to $q - 1$.
- X sends the residue $x = h^e \pmod p$ to Y.
- Y sends the residue $y = h^f \pmod p$ to X.
- X calculates $k = h^{fe} = y^e \pmod p$.
- Y calculates $k = h^{ef} = x^f \pmod p$.
- k is the mutually agreed key.

In the semi-static variant of Diffie-Hellman key exchange, one of the two, say participant Y(ollanda), has a public key $(p, q, h, b = h^a)$ and a private key $f = a$, but participant X(avier) does not.

4.5.5 ECDSA Signature

We have already transferred the Diffie-Hellman key exchange as ECDH (Sect. 3.7) as well as the ElGamal cipher (Sect. 3.8) to elliptic curves. DSA also has a variant based on elliptic curves, which is called **ECDSA** (Elliptic Curve Digital Signature Algorithm). Let p be a prime number and $y^2 = x^3 + r \bullet x + s$ an elliptic curve modulo p. Furthermore, let G be a base point on the curve with the highest possible order o. Furthermore, let q be a prime number that divides o. Then for the point $H = (o/q) \bullet G$, we have $q \bullet H = q \bullet (o/q) \bullet G = o \bullet G = O$, and H therefore has order q. Participant Y(ollanda) chooses a natural number b in the range from 2 to $q - 1$ and computes the point $B = b \bullet H$. Her public key is (p, r, s, H, q, B), and the private one is b.

If Y wants to send the message m to X(avier), she first chooses a random natural number k in the range from 2 to $q - 1$, computes $k \bullet H = (x_0, y_0)$ with remainders x_0 and y_0 modulo p, and determines $u = x_0 \pmod q$. If $u = 0$, she chooses a different k. Since q is a prime number and hence coprime with k, she uses the extended Euclidean algorithm to determine a natural number x with $x \bullet k = 1 \pmod q$, i.e., $k^{-1} = x \pmod q$. Further, Y computes the residue $v = k^{-1} \bullet (m + b \bullet u) \pmod q$. If $v = 0$, she starts the procedure with a new k. Finally, as a signature, she sends, together with the plaintext m, the residues $u = x_0 \pmod q$ and $v = k^{-1} \bullet (m + b \bullet u) \pmod q$.

To verify the signature values, receiver X computes a natural number y with $1 = y \bullet v \pmod q$ using the extended Euclidean algorithm, so that $v^{-1} = y \pmod q$ holds. This is possible because q is part of the public key of Y and because v is not 0 and therefore is coprime with the prime number q. Finally, he computes the curve point $A = v^{-1} \bullet (m \bullet H + u \bullet B) = (x_1, y_1)$ with residues x_1 and y_1 modulo p. If A is the point O, then X does not accept the signature. Otherwise, he determines $t = x_1 \pmod q$ and then considers the signature as verified if $t = u$. Indeed, if the signature really originates from Y, then for her private key b and because of $q \bullet H = O$, it follows first $A = v^{-1} \bullet (m \bullet H + u \bullet B) = v^{-1} \bullet (m + u \bullet b) \bullet H = k \bullet H$ and hence $t = u$.

Of course, again in reality a hash value of m is signed. ECDSA is a formal translation of the DSA procedure, whereby the role of h and b is taken over by the points H and B.

4.5.6 Security of the DSA and ECDSA Signature

For security reasons, the BSI guideline [BSI1] recommends key lengths for p and q in the order of 2000 and 250 bits, respectively, for the DSA signature and also for Diffie-Hellman with non-generating element. However, with increasing computer performance, the BSI recommends using prime numbers p with a length of 3000 bits for a deployment period beyond 2022. For ECDSA, a key length for q of at least 250 bits is recommended.

4.5.7 EC Standards sepc256kl and brainpoolP256r1

In standard procedures based on ECDH or ECDSA, the parameters (p, r, s, H, q) are predefined and are thus effectively part of the algorithm. Thus, only B is the public key and b is the private key. In addition to the standard P-256 (Sect. 3.7), we also want to explicitly list the two standards sepc256k1 and brainpoolP256r1 with their parameters (p, r, s, G, o). For all above mentioned standards G = H is a base point of prime order o = q.

The EC standard **sepc256k1** was proposed by the **SECG** (Standards for Efficient Cryptography Group) [BeL, SEC]:

- prime number p
  - p = FFFFFFFF FFFFFFFF FFFFFFFF FFFFFFFF FFFFFFFF FFFFFFFF FFFFFFFE FFFFFC2F
- Elliptic curve $y^2 = x^3 + 7$ (i.e. r = 0 and s = 7)
- Base point $G = (x_G, y_G)$ of prime order o
  - x_G = 79BE667E F9DCBBAC 55A06295 CE870B07 029BFCDB 2DCE28D9 59F2815B 16F81798
  - y_G = 483ADA77 26A3C465 5DA4FBFC 0E1108A8 FD17B448 A6855419 9C47D08F FB10D4B8
  - o = FFFFFFFF FFFFFFFF FFFFFFFF FFFFFFFE BAAEDCE6 AF48A03B BFD25E8C D0364141

BSI recommends the EC standard **brainpoolP256r1** [BSI1, LoM]:

- prime number p
  - p = A9FB57DB A1EEA9BC 3E660A90 9D838D72 6E3BF623 D5262028 2013481D 1F6E5377
- Elliptic curve $y^2 = x^3 + r \cdot x + s$
  - r = 7D5A0975 FC2C3057 EEF67530 417AFFE7 FB8055C1 26DC5C6C E94A4B44 F330B5D9
  - s = 26DC5C6C E94A4B44 F330B5D9 BBD77CBF 95841629 5CF7E1CE 6BCCDC18 FF8C07B6
- Base point $G = (x_G, y_G)$ of prime order o
  - x_G = 8BD2AEB9 CB7E57CB 2C4B482F FC81B7AF B9DE27E1 E3BD23C2 3A4453BD 9ACE3262
  - y_G = 547EF835 C3DAC4FD 97F8461A 14611DC9 C2774513 2DED8E54 5C1D54C7 2F046997
  - o = A9FB57DB A1EEA9BC 3E660A90 9D838D71 8C397AA3 B561A6F7 901E0E82 974856A7

The parameters are represented hexadecimally, whereby 4 bits are combined into numbers from 0 to 15. The letters stand for the two-digit numbers A = 10, B = 11,…, F = 15.

4.5.8 EC Standard Curve25519

So far we have only encountered elliptic curves in the so-called Weierstrass form $y^2 = x^3 + r \cdot x + s$. The EC standard **Curve25519**, on the other hand, which has not yet been explicitly mentioned, is a different representation of elliptic curves, in this case $y^2 = x^3 + 486{,}662 \cdot x^2 + x$ modulo the prime $p = 2^{255} - 19$. The curve Curve25519 was proposed in 2005 by **Daniel Bernstein** (born 1971). Unlike the usual form of elliptic curves, it allows the use of algorithms immune to so-called side-channel attacks. Here, it is not the cryptographic cipher itself that is attacked, but a specific implementation in a device (e.g. a chip card) [WPC29].

4.6 Online Banking

We would now like to discuss the data security of **online banking**. Online banking is the processing of banking transactions with the help of laptops, tablets or smartphones, where you can access the bank computer directly from home or on the road via the Internet. Online banking can be used to make account inquiries, but also to make transfers.

4.6.1 HBCI and FinTS

HBCI (Homebanking Computer Interface) is a German standard for online banking, adopted by the German banking industry in the 1990s. In 2002, HBCI was renamed **FinTS** (Financial Transaction Services) with version 3.0. This standard defines the transmission protocols, message formats and security procedures, but the user interface is not standardised. Each bank can design this according to its "corporate identity". The customer first authenticates himself with user name and PIN on the Internet at the online banking of his bank. After this login, all informative business transactions are available to him, for example, the inquiry of account balances.

4.6.2 2-Factor Authentication with TAN

With the conventional security concept, the so-called **2-factor authentication**, a new **TAN** (transaction number) is additionally required for each account-moving transaction, such as transfers, which is sent to the customer by the bank via a separate channel.

With the so-called **TAN list**, the customer receives a list of different TANs from his bank by post, one of which must be entered for each transfer order at the online banking. Once a TAN has been used, it has expired and can no longer be used. As a moderate further development, indexed TAN lists are used (so-called **iTAN**), in which the TANs are numbered consecutively. As part of the online transfer order, the customer is requested to enter the TAN corresponding to the number displayed. However, the use of TAN list and the iTAN procedure has expired since the end of 2019.

A still common procedure is called **mTAN** (mobile TAN). After secure Internet transmission of the transfer completed at the online banking, the bank sends the customer a TAN that can only be used for this transaction together with (parts of) the target account number by SMS to his mobile phone. The transfer order must then be confirmed with this TAN within a few minutes before it is actually executed. In another variant, the **eTAN** procedure, the customer uses a TAN generator. This generates a TAN based on a displayed BAR code and the inserted bank card of the customer as well as the current date and time, which is also only valid for a short period of time. The bank can use the same algorithm to check the TAN entered.

The use of the TAN list and iTAN at least ensures that a passive attacker cannot repeatedly use the TAN once it has been intercepted. But why have these procedures been phased out? The reason is that they are vulnerable to a man-in-the-middle attack. These attacks are carried out via Trojans on the customer's device. The scenario then looks like this:

- The fraudster foists an Internet address on his victim's device that leads to a fictitious page and makes the customer believe that he has official access to his bank's online banking. The customer falls for the trick and, by entering his access data, makes them available to the fraudster.

- The fraudster uses the access data and simultaneously establishes a connection to the bank's real online banking.
- Meanwhile, the customer enters the data for a bank transfer in the fake system.
- The fraudster also starts a transfer process from the customer's account at the real online banking, but with a large amount to an account held in the Cayman Islands. In the last step, the fraudster is asked to enter the TAN (for the displayed number) for authorization.
- The fraudster then asks the customer in the fake system to enter the TAN (for this number).
- The customer enters the correct TAN from his list, the fraudster receives it, enters it as part of his own ongoing transfer process and successfully authorises the transfer in this way.

If two separate end devices are used for the mTAN procedure, e.g. a laptop for the transfer and a smartphone for SMS, this can be considered sufficiently secure, as the probability that both end devices are infected is low. However, the whole thing becomes critical if the fraudster can hack into both communications via a man-in-the-middle attack and also change the content of the SMS. The 2-factor authentication is just only a doubled verification based on the static PIN and the dynamic TAN.

4.6.3 FinTS with Digital Signature

To use the more modern security concept RAH-7 and RAH-9 available with FinTS, the customer needs a special individual chip card (ICC), which is delivered by the bank in a secure way, and a chip card reader, which must be connected to the customer's computer. For the time being, however, the RAH-10 software solution installed on the customer's computer itself is also possible. Figure 4.11 visualises the process of a bank transfer with FinTS and digital signature described below:

- To initiate a bank transfer, the customer enters the online banking of his bank on the Internet, logs in with user name and PIN and prepares the transfer there.
- He then inserts his chip card into the card reader and enters his PIN again for security. Thereupon

Online Banking with PIN	Card Reader	Chip Card	Internet with TLS	Bank Server	Bank Server
Customer Enters Bank Transfer Data	Customer Inserts Bank Card (Chip Card)	Signs and Encrypts Bank Transfer Data	Transmits Bank Transfer	Decrypts Bank Transfer and Checks Signature	Executes Transfer Order if Check is Successful

Fig. 4.11 Schematic workflow for online banking with digital signature

  - the digital signature of a digital fingerprint of the transfer is made using the private key of the customer stored on the chip card,
  - a key is randomly generated and the entire transfer is encrypted using a symmetric cipher, and
  - this key is in turn encrypted with the public key of the bank and attached.
- The transfer is then sent in this form TLS-secured via the Internet to the bank server.
- As soon as the bank receives the transfer, it decrypts it and checks the signature using the customer's public key.
- Only if this check is successful the transfer order will actually be executed.

This procedure is not only tap-proof, but also tamper-proof against a man-in-the-middle attack. However, the FinTS standard still allows the mTAN and eTAN procedures as alternatives.

4.6.4 Encryption and Signature Procedures for FinTS

We now want to list the procedures that are used in FinTS [BDB]. While in version 3.0 Triple-DES was allowed as an alternative symmetric cipher for encrypting the transfer data, in version 4.0 only AES with a key length of 256 bits is allowed. The key is generated by generating a 256-bit random number. AES encryption takes place in CBC operating mode.

The public key cipher both for encrypting the AES key and for signing transfer data is RSA. The customer keys are generated by the processor on the individual chip card (ICC). For this purpose, prime numbers p and q are generated and the RSA module $n = p \bullet q$ is calculated based on the currently recommended bit length of about 2000 bits. The second public key component e is fixed as the prime number $e = 2^{16} + 1$, while the private key d is computed using the extended Euclidean algorithm from $1 = d \bullet e + b \bullet (p - 1) \bullet (q - 1)$. As a rule, when the customer accesses FinTS for the first time, the public keys of the customer and the bank are mutually exchanged. Alternatively, this is also possible on data carriers. The private customer key d is stored in an area of the chip from which it cannot be read and therefore never leaves the chip card.

The hash function SHA-2-256 is used to generate the digital fingerprint of the transfer data. In addition, the transfer data is compressed using the zip format before being sent.

4.7 Blind Signature and Cryptocurrencies

4.7.1 Anonymity and Blind Signature

In various cryptographic applications, the participants sometimes wish to remain unrecognized, i.e. anonymous. We therefore also want to describe an anonymized form of digital

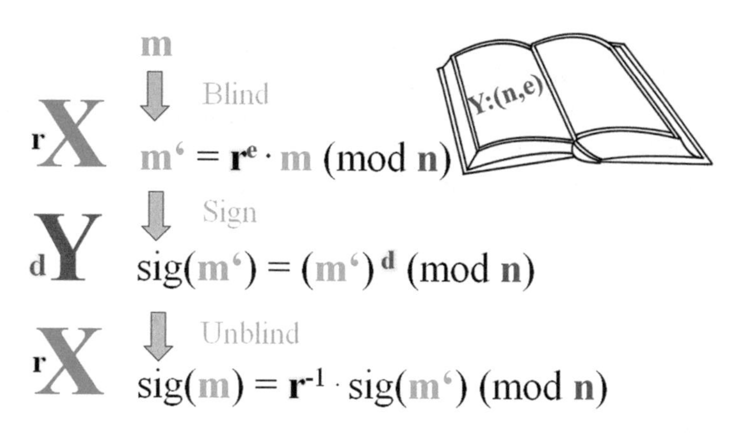

Fig. 4.12 Blind signature

signature, the so-called **blind signature**. Here, X(avier) acquires a valid digital signature from Y(ollanda) for an information content that Y cannot even recognize when signing. Such a method was first proposed in 1981 by **David Chaum** (born 1955) based on the RSA cipher. For this, let (n, e) be Y's RSA public key and d be her private key. X chooses a random natural number r in the range from 2 to n – 1, which is coprime with n, and keeps it secret. Further, X computes the value $r^{-1} = x \pmod{n}$ by using the extended Euclidean algorithm to determine the multiple sum $1 = x \bullet r + y \bullet n$. Using the random number r, X modifies his text m to be signed to $m' = r^e \cdot m \pmod{n}$ and submits the "blinded" m′ to Y for digital signature. Since r was chosen at random, Y cannot infer back to m. "Faithfully" Y signs the blinded text m′ with her private key d, thus calculating $sig(m') = (m')^d (\mathrm{mod}\ n)$ and submits this signature to X. The latter multiplies Y's signature by r^{-1} modulo n, thereby obtaining $r^{-1} \cdot sig(m') = r^{-1} \cdot (m')^d = (r^{-1} \cdot (r^e \cdot m)^d) = (r^{-1} \cdot r^{ed} \cdot m^d) = (r^{-1} \cdot r \cdot m^d) = m^d = sig(m) \pmod{n}$ and in this way has "unblinded" the blinded signature again. Therefore, X indeed has the digital signature $sig(m) = r^{-1} \cdot sig(m') (\mathrm{mod}\ n)$ of Y for the original text m. Everyone can verify this by using the public key (e, n) of Y and checking whether $(sig(m))^e = (m^d)^e = m \pmod{n}$ really holds. Figure 4.12 illustrates the procedure graphically.

4.7.2 Cryptocurrency eCash

But who would blindly sign "faithfully" in Chaum's procedure? Yes, there are such situations. Chaum has used his concept of the blind signature to define a so-called **cryptocurrency**. This is understood to be a new manifestation of money, in addition to the classical banknotes and coins of the central bank and the book money of commercial banks. A cryptocurrency is held in digitally encrypted form, stored on a server or computers somewhere in the network or in a cloud. In addition to more practical requirements such as user-friendliness and availability, from a cryptographic point of view, cryptocurrencies are required to be counterfeit-proof and verifiable. If one adds the request to be anonymous here, there can no longer even be a kind of serial number as with banknotes, but it is then more of a digital coin system.

Chaum conceived of his cryptocurrency, which he called **eCash,** as a digitally stored claim on a financial institution. So in his model, banks issue eCash shares, which customers can purchase with "normal" money to make purchases. For a piece of eCash worth say US$100, Bank B issues the following specifications:

- a specially created public RSA key (n, e), but keeps the associated private key d secret, and
- a redundancy scheme according to which digital data sets must be prepared.

As a simple example, such a redundancy scheme could look like this: the data record must consist of a digital string of length 5, which must then be repeated twice, e.g. 11001 11001 11001. In order to purchase a piece of eCash worth US$100 from bank B, customer C has bank B blindly sign a digital string m prepared according to the redundancy scheme with the RSA key. The bank then debits the US$100 from his account. The value $g = sig(m) = m^d$ (mod n) is then the piece of eCash worth US$100 that customer C purchased from bank B. It is clear that g as a digital signature could only be generated using Bank B's private key d (counterfeit-proof) and that anyone can verify the authenticity of g using Bank B's public key (n, e) and the published redundancy scheme (verifiable). Moreover, g does not reveal any information about C, not even Bank B knows that it actually signed m and thus issued the piece of eCash g = sig(m) to C (anonymous).

This is a brief sketch of the theory, and we do not want to go into any further details of the technical and organisational implementation of eCash. For the commercial marketing of his financial product, Chaum founded the company DigiCash at the beginning of the 1990s, which was able to win Deutsche Bank and Credit Suisse, among others, as European eCash licensees. However, Chaum was probably too far ahead of his time with the cryptocurrency eCash. In any case, DigiCash went bankrupt at the end of the 1990s.

4.7.3 Cryptocurrency Bitcoin

There are now many different cryptocurrencies on the market, which are usually much more complicated in structure and implementation than the comparatively simple eCash. Probably the best known, namely **Bitcoin (BTC)**, was invented in 2009 under the pseudonym **Satoshi Nakamoto** and traded publicly for the first time. Bitcoin currently has a market share of over 50%. It is followed by the cryptocurrencies **Ripple/XRP** and **Ether/ETH**. The conversion rate of Bitcoins into other means of payment is determined by supply and demand. Thus, unlike eCash, Bitcoin does not require commercial banks.

Bitcoins are exchanged electronically between the parties involved in the trade. A digital signature scheme is required for authentication. Bitcoin uses ECDSA with the standard sepc256k1 (Sect. 4.5) and the hash function SHA-2-256 (Sect. 4.3). Each participant needs an ECDSA public key and a private key. The public key also serves as the basis for establishing the identity of the participant and thus, as a random-looking string, simultaneously guarantees the desired anonymity. Each participant must import their private key into their so-called **Bitcoin wallet** and then store it securely. In addition, the Bitcoin wallet also holds his current account balance, so it must be protected with a strong password.

Bitcoin deliberately dispenses with any central authority (such as central banks or commercial banks) that could mediate financial transactions; instead, these are to take place directly between the participants involved. However, if data were stored in the cloud or on the Internet, one would have to access at least one or even several servers. Therefore, Bitcoin data storage is decentralized in a so-called **P2P network** (peer-to-peer), where each participant is directly connected to others. The prerequisite for this is the installation of free software.

4.7.4 Bitcoin Transactions

A **Bitcoin transaction** involves the instruction of a certain equivalent amount of Bitcoins from the public key B_A of the instructing party to the public key B_E of the recipient. However, it is not a specific amount of Bitcoins that is transferred from the instructing party's Bitcoin wallet, but a previous incoming transaction to the instructing party's public key B_A is transferred to the public key B_E of the new recipient. This new transaction has exactly the equivalent value of bitcoins that the previous transaction had. However, since there is usually no previous transaction available with exactly the equivalent value that is now to be transferred, Bitcoin transactions usually have at least two instructions. One is sent to the public key B_E of the actual recipient and the other is sent to the public key B_A of the instructing party, who then transfers the remaining money back to himself. With Bitcoin, there are therefore no accounts in the actual sense that have a credit balance. The account balance in a participant's Bitcoin wallet results from the transactions received to the participant's public key that have not yet been transferred on.

When a transaction is completed, the hash value of the transaction data is digitally signed by the instructing party using his private key b_A via ECDSA, the signature is attached to the transaction and it is thus protected against changes. To authenticate a transaction, it is transmitted to the P2P network and thus made available to all participants connected to the instructing party. These verify the signature with the instructing party's public key B_A and thus check whether the transaction is valid. They then forward the transaction to other participants in the P2P network.

4.7.5 The Bitcoin Blockchain

The **Bitcoin blockchain** consists of a chain of blocks, each of which contains a certain number of Bitcoin transactions. New blocks are created in a computationally intensive

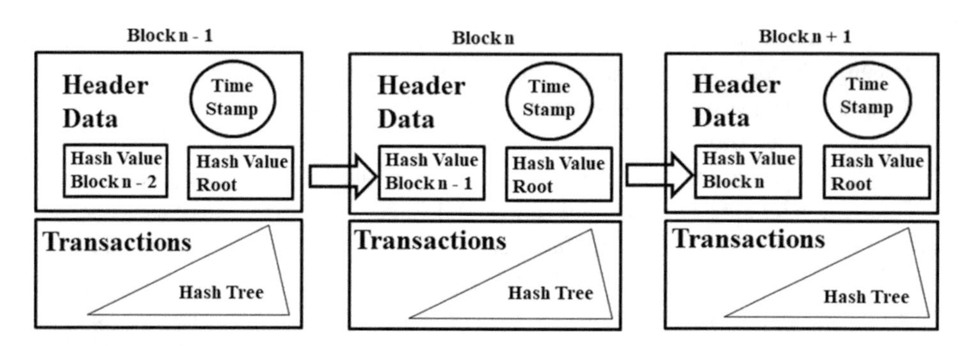

Fig. 4.13 Bitcoin blockchain

process called **mining**, and then distributed to participants via the Bitcoin P2P network. Mining, meanwhile, consumes large amounts of energy.

Within a Bitcoin block, two transactions are combined in pairs and a common hash value is calculated for their individual hash values. The resulting hash values are again combined in pairs and a common hash value is calculated for each. In this way, a tree structure of hash values is successively created, the last of which, the so-called **root hash,** is stored in the block header together with a time stamp. The blocks are linked to each other using the hash values of their header data. This is done in such a way that the hash value of the header data of the previous block is also written to the header of each block. The procedure is shown schematically in Fig. 4.13. Thus the sequence of the blocks is clearly defined. In addition, the subsequent modification or even deletion of previous blocks or transactions is practically excluded, since the hash values of all subsequent blocks would also have to be recalculated in a short time.

4.8 Password Security and Challenge Response

4.8.1 Password as Hash Value

Secure communication is generally preceded by authentication of the users with a verifier (Sect. 4.1). We have seen this in the example of PIN entry for online banking (Sect. 4.6), but also before sending an e-mail (Sect. 4.4), you have to log on to a computer network with your password. Passwords are usually used for servers or laptops with hard disks, whereas PINs are used for chip cards (ICC). We will look in more detail at how authentication works for mobile phones and credit card purchases (Sect. 4.9). However, we will first look at the issue of password security in general. A personal password or PIN is personal knowledge that must never be passed on or disclosed. However, if you enter a password or PIN when logging in, it must be possible for the verifier to check the entry. Therefore, a file must be stored on the hard disk of your own laptop, on a chip card or on a server in the Internet, which contains the user name and the corresponding password or PIN. However, at least the system administrator has authorized access to this, but possibly also other unauthorized attackers. The solution to this problem is once again a cryptographic hash function, i.e. the password or PIN is only stored in the form of its hash value. Under the SHA-1 hash function, for example, two popular passwords look like this [Rau]:

```
admin1234 7B902E6FF1DB9F560443F2048974FD7D386975B0
password  5BAA61E4C9B93F3F0682250B6CF8331B7EE68FD8
```

The 160-bit string of SHA-1 is again represented here in hexadecimal. Here, 4 bits are combined into numbers from 0 to 15, whereby the two-digit hexadecimal numbers are written as letters, namely A = 10, B = 11,..., F = 15.

Of course, with hash values of passwords the focus does not lie on data compression, as is the case with digital signatures. Passwords are usually only a few characters long. Instead, other properties of cryptographic hash functions are used. For example, every hash value of a given hash function has the same length, no matter how long the corresponding password is. Thus, one cannot infer the length of the password. Also, the hash value does not allow one to infer the number of digits and special characters. Hash functions cause great confusion and diffusion, a property inherited from block ciphers as their building blocks. Hash values for similar passwords therefore differ significantly; even small changes result in fundamentally different hash values. Here is an example using the hash function SHA-2-224 [WPSH2e], where just a period has been added

```
The quick brown fox jumps over the lazy dog
730e109bd7a8a32b1cb9d9a09aa2325d2430587ddbc0c38bad911525

The quick brown fox jumps over the lazy dog.
619cba8e8e05826e9b8c519c0a5c68f4fb653e8a3d8aa04bb2c8cd4c
```

Overall, it must not be possible to infer an associated password from a given hash value using efficient methods. And exactly this is formally guaranteed by the one-way property of a cryptographic hash function.

4.8.2 Attacks on Passwords

Nevertheless, it is well known that attempts are always made to crack password files. A brute-force attack is used to try out all theoretically conceivable passwords in sequence and compare their hash value with the stored values. With about 70 characters, consisting of upper and lower case letters, digits and some special characters, there are exactly 70^8 different passwords of length 8, i.e. a little less than 10^{15} or 2^{50} ones. This is in the order of DES keys and is therefore no longer a problem for computers today. An even faster way to achieve the goal is to use a **dictionary attack** to go through a dictionary instead of arbitrary passwords, supplemented by first names and calendar data. There are also so-called **rainbow tables** with a specially developed data structure that allows an extremely fast search for passwords for a given hash value.

4.8.3 Salted Hash Values

You can increase the password security a bit more by using the concept of **salting**. This adds a little "salt to the corned beef hash". Each time a password is entered, a few characters that are as meaningless as possible are automatically added, such as &7T?a$. This makes the hash look completely different and avoids passwords that are too simple. Salting can also be customized with individual additions for each user. This leads to the fact that for two users with the same password nevertheless different hash values are stored. Another aspect is also advantageous: Many users use the same password for some or all of their applications. However, since they all use different salting, the hash values look completely different everywhere.

4.8.4 RADIUS Server for WLAN and DSL

A **RADIUS** server (Remote Authentication Dial-In User Service) is a central server in a computer network that takes over the authentication of a user by checking the user name and password when dialing in. In this way, all settings can be managed centrally, so that the access data of the users are available everywhere and at any time. RADIUS with the so-called **EAP** protocol (Extensible Authentication Protocol) is the de facto standard for authentication in larger WLAN installations and for DSL on the Internet.

4.8.5 Challenge-Response Authentication

All the effort involved in storing passwords is only necessary because the verifier must be able to check the subscriber's knowledge in the form of his password. Another weakness of the method is also that an attacker who has intercepted a password does not necessarily have to use it immediately, but can use it sometime later and even multiple times.

These problems can be circumvented with **challenge-response authentication**. User T(ina) is confronted with a challenge by verifier V(ictor), which she can only solve based on her secret knowledge. In most cases, this is a kind of "arithmetic problem". User T sends the solution as a response back to the verifier V, who checks the answer. If it is correct, user T has been successfully authenticated by verifier V. Necessary in this procedure is that the task is randomly generated and thus varies sufficiently with each new authentication process. Furthermore, it must be possible for the verifier V to check the correctness of the answer without knowing the user T's secret knowledge.

4.8.6 Challenge-Response with Digital Signature

One way to implement challenge-response authentication is again the digital signature sig(.). Thus, if user T(ina) wants to authenticate herself to verifier V(ictor), V sends a random number m to T as a challenge. T, in turn, uses her private signature key to digitally sign the random number m. She then sends sig(m) as a response back to V. The latter verifies the digital signature with T's public key. If this verification returns an "o. k.", then T has successfully authenticated herself to V. Figure 4.14 visualizes the process.

While passwords can usually be entered by hand, authentication with a digital signature requires the use of a chip card (ICC) due to the complexity of the calculation. Formally, the cardholder authenticates himself by possession and the card itself by knowledge towards the verifier.

At first glance, verifier V(ictor) does not learn what the secret is in the form of the private key of user T(ina) during challenge-response authentication with a digital signature. However, V could obtain more and more partial information about the private key by suitably strategic choice of the random number m as well as by repeated application of the procedure in each authentication process of T. This can be prevented by making the task much more complex, limiting the variation possibilities on the part of the verifier V, but making those of the participant more flexible (Sect. 4.9). However, there are also procedures in which it can be proved that absolutely no information flows, no matter which strategy V uses. This is called **zero-knowledge authentication**, and probably the best-known procedure is that of **Amos Fiat** (b. 1956) and **Adi Shamir** (b. 1952) in 1986. Similar to the Rabin cipher, it relies on "square roots" modulo a large number $n = p \cdot q$ [BNS, Buc].

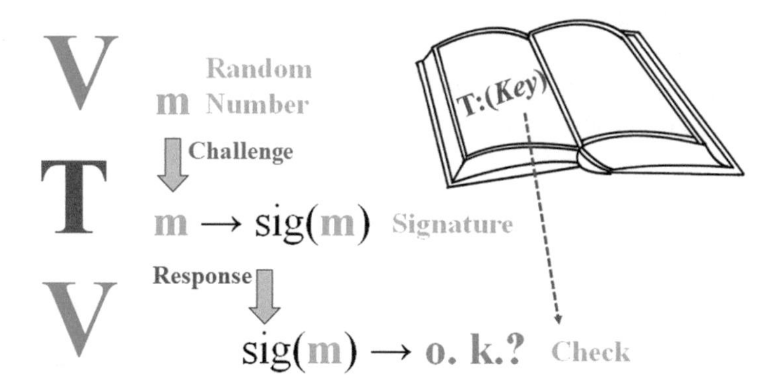

Fig. 4.14 Challenge-response authentication with digital signature

4.9 Mobile Phone, Credit Card and Passport

Finally, we will now look at some examples of how the authentication of users is implemented in practice.

4.9.1 Authentication in GSM Mobile Communications

We have already explained the encryption method used in 2. generation GSM mobile communications (Sect. 2.3). It uses personalised chip cards (ICC). These so-called **SIM cards** (Subscriber Identification Module) are issued by the network operators to their customers. Each subscriber is assigned a 128-bit subscriber authentication key k_i, which is stored on the SIM card in the mobile phone and in the mobile communications server and which we will now refer to as k for short.

Now we will deal with the authentication of mobile communications subscribers. The subscriber authenticates himself by knowledge, namely the PIN, and possession, namely the SIM card. The PIN entry on the mobile phone is checked by the chip of the SIM card. However, it can also be deactivated. If the PIN is entered incorrectly three times in a row, the SIM card is automatically blocked. To unlock it again, the **PUK** (Personal Unblocking Key) is required.

After successful PIN entry, the SIM card is authenticated by the network operator's mobile communications server on the basis of its knowledge, namely the subscriber key k. The so-called **A3** algorithm is used for this purpose. Similar to the A8 algorithm for key generation, the definition of A3 is also the responsibility of the respective network operator; it is also kept secret as far as possible. In any case, the mobile communications server sends a 128-bit random number RAND to the subscriber's mobile phone as a "challenge". The subscriber's SIM card calculates a 32-bit response SRES (signed response) from RAND and k using the A3 algorithm and sends it back to the mobile communications server as a response. There, the subscriber's individual key k is read from a database and SRES is also calculated. Only if the two values match is the SIM card authenticated and the subscriber is granted access to the network. Figure 4.15 visualizes the process.

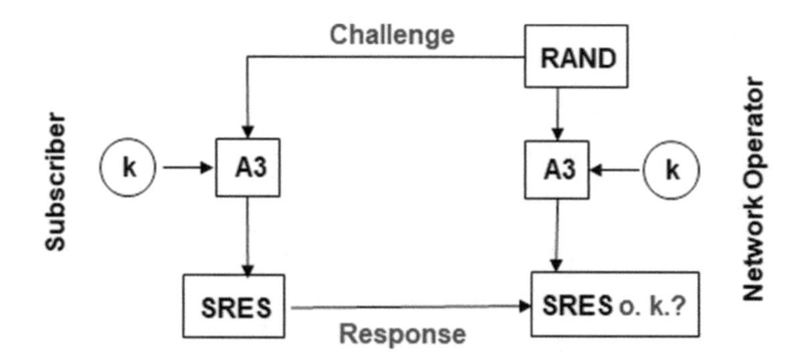

Fig. 4.15 Authentication of mobile communications subscribers with GSM

4.9.2 Authentication in UMTS/LTE Mobile Communications

In the UMTS and LTE standards of the 3. and 4. generation, a similar challenge-response procedure is used to authenticate a mobile communications subscriber, also using the 128-bit subscriber key k and the 128-bit random number RAND. However, the A3 algorithm of GSM is replaced by a standardized procedure, which still leaves some possibilities for the network operator to configure suitably. The entire procedure is called **MILENAGE**, of which the algorithm for authentication is only a part. Among other things, MILENAGE also generates a 128-bit cipher key which, together with the A5/4 cipher (Sect. 2.7), encrypts the data to be transmitted, such as the calls or Internet pages. We explain the authentication part of the algorithm using the workflow in Fig. 4.16 [ETSI3].

- The input value to the MILENAGE procedure is the 128-bit random number RAND.
- $E_k = E(\bullet, k)$ is a block cipher on 128-bit blocks, which depends on a 128-bit key k. The block cipher is applied multiple times in the procedure. The standard leaves the choice of E_k open in principle, but strongly recommends using AES with key length 128 bits.
- OP is a 128-bit constant that can be freely configured by the respective network operator. It is modified to OP_C using the block cipher E_k and thus added several times bit by bit $\oplus$ within the procedure.

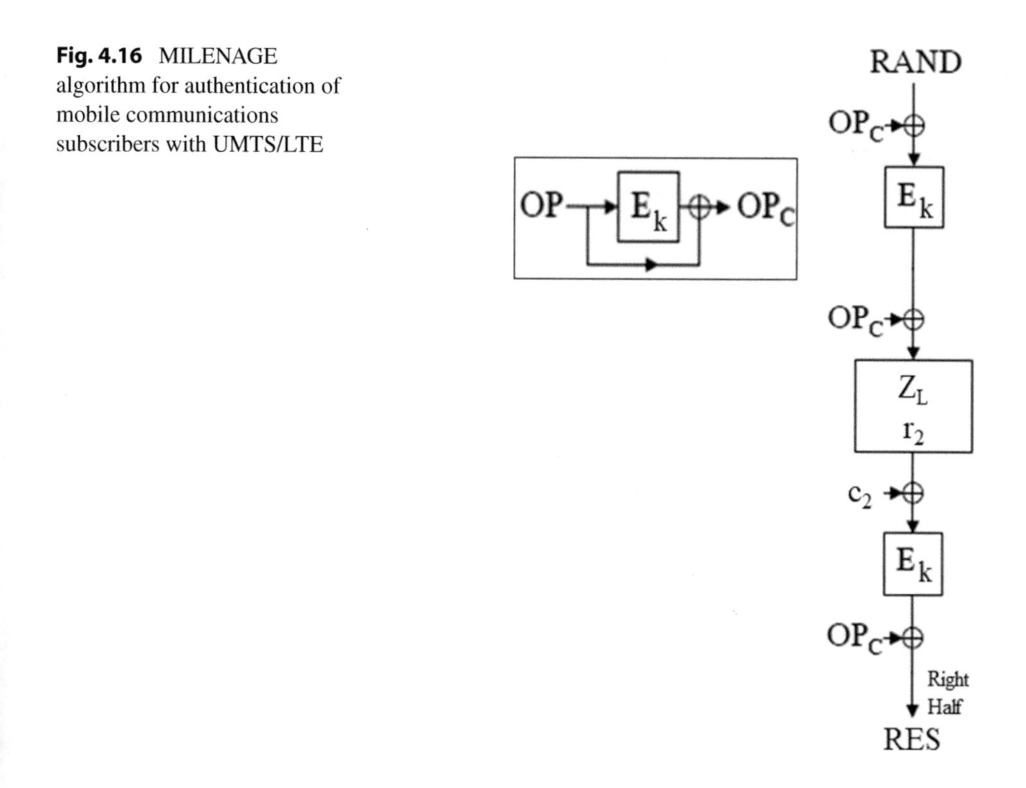

Fig. 4.16 MILENAGE algorithm for authentication of mobile communications subscribers with UMTS/LTE

- For the 128-bit constant c_2, which is added once bitwise $\oplus$ in the procedure, the standard contains $c_2 = 00\ldots001$ as a suggestion.
- For the constant r_2, which can assume the values 0, 1,..., 127, $r_2 = 0$ is proposed in the standard. In principle, it causes the cyclic shift Z_L of a bit string by r_2 positions to the left.
- The output value of the MILENAGE procedure is initially a 128-bit number. From this, the 64-bit number RES is derived as the right half, which is used for the authentication of the SIM card and thus of the subscriber.

The GSM/UMTS/LTE standard does not specify how often authentication is to be performed. It must be performed at least when the mobile phone is switched on, but can also be performed operator-dependently when dialling into a new cell tower and automatically in fixed time cycles.

4.9.3 Credit Card and Secure Data Transmission

The **EMV** (Europay International, MasterCard and VISA) takes care of the creation and review of specifications and requirements for secure payment with **credit cards**. Today, it is a joint organization of American Express, Discover, JCB, Mastercard, UnionPay and Visa in cooperation with numerous banks, retailers and industry [EMV1].

Of course, in the field of credit card payments, secure data transmission is of particular importance, especially in view of the many parties involved in the process. This involves

- the holders of credit cards,
- the issuers of credit cards (e.g. Master, Visa, American Express) with which the holders have a contract,
- the merchants who accept credit cards for purchases,
- the holders' principal banks, which ultimately transfer the credit volume to the issuer, and
- the operators of the credit card terminals, which ensure smooth payment transactions between the cardholder, the merchant and the credit card issuer.

The specification and guideline issued by EMV [EMV2, EMV4] preferably permits Triple-DES for encryption, but also AES with key lengths of 128, 192 and 256 bits, operated in ECB or CBC mode. A CBC-MAC is recommended for authentication of the data transmission (Sect. 4.1).

4.9.4 Authentication When Purchasing with a Credit Card

Here we want to look more specifically at the authentication of cardholders when making purchases with credit cards. Every credit card is an ICC (Integrated Circuit Card) and therefore contains an integrated chip. The chip stores the cardholder's individual public RSA key (n, e) and his private d, both of which were generated when the card was created. For e, only the values 3 and $2^{16} + 1$ are permitted. The private key d is stored in an area of the chip from which it cannot be read.

The holder of a credit card authenticates himself by knowledge (PIN) and possession (credit card) when reading into a terminal. In the process, the PIN entry is verified by the chip of the credit card. The chip in turn is authenticated by the terminal on the basis of knowledge, as we will explain in principle using **DDA** (Dynamic Data Authentication) [EMV2, EMV3].

The terminal generates a bit string of defined terminal data i_T and prefixes it with a random number z_T of 4 bytes length. It sends the "challenge" $z_T||i_T$ to the chip of the credit card.

The chip supplements the bit string with defined data i_C from its memory and prefixes it with a further 2 to 8 byte random number z_C. This results in the bit string $m = z_C||i_C||z_T||i_T$, where we have neglected a few more format bits here. In any case, the chip computes a 160-bit hash value h(m) using the hash function SHA-1. We now use $m_0 = z_C||i_C$ as an abbreviation. The procedure takes care that the bit length of $m_0||h(m)$ is smaller than the one of the RSA module n of the cardholder. Now the chip signs the bit string $m_0||h(m)$ with the private key d of the cardholder and therefore sends $sig(m_0||h(m)) = (m_0||h(m)^d \pmod n)$ as "response" to the terminal.

This uses the public key (n, e) of the cardholder and calculates $m_0' h' = \left(sig\left(m_0 h\left(m\right)\right)\right)^e = \left(m_0 h\left(m\right)\right)^{de} \left(\bmod n\right)$. Here h' denotes the 160 rightmost bits in the calculated bit string and m_0' the rest. If the signature is correct, this should result in $m_0' = m_0$ and $h' = h(m)$. The terminal therefore interprets m_0' as the bit string $m_0 = z_C||i_C$ of the chip, which is unknown to it, and supplements the bit string $z_T||i_T$, which is known to it, to $m' = m_0'||z_T||i_T$.. From m' it calculates the hash value $h(m')$. If this is equal to the received h', the terminal considers the chip and thus the credit card as authenticated. For only a signature with the correct private key d could result in this coincidence of the hash values.

4.9.5 Electronic Passport ePassport

The **ePassport** (electronic passport) was introduced in Germany in 2005. Originally, there was a chip in its cover with which a terminal can exchange data contactlessly via **RF** (Radio Frequency). Since 2017, the chip has been integrated into the passport's chip card. The chip stores the personal data as well as the biometric picture and two fingerprints of the passport holder [BSI4].

However, the chip of the ePassport also contains both the public (p, q, h, b) and the private DSA key a of the passport holder (Sect. 4.5). The private key is located in an area of the chip from which it cannot be read. Therefore, even if the complete chip is "cloned", it is not possible to copy the private key as well. The public key of the passport holder is readable, but again secured by a digital signature of the issuing authority. In the passport creation phase, a hash value of the personal data stored on the chip is also digitally signed and stored with the passport holder's DSA key.

4.9.6 Authentication on Entry or Exit with ePassport

The passport holder authenticates himself at the automatic passport control by possession (ePassport) and by his characteristics (biometric picture, fingerprint). In the process, the characteristics are verified by the chip integrated in the ePassport. The chip is in turn authenticated by the terminal by checking its knowledge. This is done in two steps, which we will now explain in principle [BSI5], [BSI6].

Passive authentication (PA) is used to verify the authenticity of the passport and the integrity of the data on the chip. To do this, the terminal reads the personal data and their digital DSA signature from the chip and verifies the signature with the passport holder's public key. However, with passive authentication, copying the data from one chip to another would remain undetected.

Chip authentication (CA2) additionally is used to recognize "cloned" chips in ePassports. For this purpose, the terminal generates a random natural number f, reads the public key of the passport holder and sends $c = h^f \pmod{p}$ to the chip as a "challenge".

The chip uses its private key a to calculate the remainder $k = c^a = h^{fa} \pmod{p}$, chooses a random natural number r and uses a hash function h(•) to calculate the hash value $k_m = h(k||r)$, where here k||r is to be interpreted as the stringing together of the digital representations of the numbers k and r. In order to calculate the CBC-MAC of c, k_m is suitably truncated to k_m' and thus $t = mac_{CBC}(c, k_m')$ is calculated (Sect. 4.1). Finally, the chip sends the values r and t to the terminal as a "response".

The terminal, for its part, uses its random number f to calculate $k = b^f = h^{af} \pmod{p}$ from the passport holder's public key, which ultimately corresponds to a semi-static Diffie-Hellman key exchange (Sect. 4.5). Now the terminal is also able to compute the hash value $k_m = h(k||r)$ from the received r, and in turn to compute the CBC-MAC $mac_{CBC}(c, k_m')$ from it. If this matches the received value t, the terminal considers the ePassport as authenticated. Indeed, a "cloned" chip cannot have the original private key a, and if it were simply to use a different private key, the Diffie-Hellman keys k and subsequently t computed on both sides would differ. If, on the other hand, entirely new DSA keys had been generated for a "cloned" chip, this would have been noticed during passive authentication, since the public key is protected against unnoticed changes by an official digital signature.

In addition to DSA and semi-static DH, the BSI guideline also allows ECDSA and semi-static ECDH, among others with the standards P-256 and brainpool256r1. Incidentally, a key k_c is also generated from the Diffie-Hellman key k in a similar way for the encryption of data transmission. Triple-DES and AES with key lengths of 128, 192 and 256 bits are permitted as symmetric ciphers for CBC-MAC and data encryption, and SHA-1 and SHA-2 as hash functions.

Bibliography

[3GPP] 3GPP: A5/3 Encription Algorithm for GSM (Technische Spezifikation). Sophia Antipolis Valbonne/Frankreich (2003). https://www.gsma.com/aboutus/wp-content/uploads/2014/12/a53andgea3specifications.pdf

[BeL] Bernstein, D., Lange, T.: SafeCurves: Choosing Safe Curves for Elliptic-Curve Cryptography (Internet-Information). Eindhoven/Niederlande. https://safecurves.cr.yp.to/. Accessed 10 Apr 2019

[Beu] Beutelspacher, A.: Kryptologie (Sachbuch). Springer Spektrum, Wiesbaden (2015)

[BNS] Beutelspacher, A., Neumann, H., Schwarzpaul, T.: Kryptografie in Theorie und Praxis (Lehrbuch). Vieweg+Teubner, Wiesbaden (2010)

[BiC] Bitcoinworld: Bitcoin-Lexikon (Internet-Information). https://www.bitcoin-welt.com/bitcoin-lexikon/. Accessed 10 Apr 2019

[Blu] Bluetooth: Bluetooth Core Specification v. 5.0 (Technische Spezifikation). (2016). https://www.bluetooth.com/specifications/bluetooth-core-specification

[Bre] Bressoud, D.: Factorization and Primality Testing (Lehrbuch). Springer, New York (1989)

[Buc] Buchmann, J.: Einführung in die Kryptographie (Lehrbuch). Springer Spektrum, Berlin (2016)

[BSI1] Bundesamt für Sicherheit in der Informationstechnik: Kryptographische Verfahren 1: Empfehlungen und Schlüssellängen (Technische Richtlinie). Bonn/Deutschland (2018). https://www.bsi.bund.de/DE/Publikationen/TechnischeRichtlinien/tr02102/index_htm.html;jsessionid=D4F0ACAD39ED0893ECBE3F951AE6B66C.2_cid360

[BSI2] Bundesamt für Sicherheit in der Informationstechnik: Kryptographische Verfahren 2: Verwendung von Transport Layer Security (TLS) (Technische Richtlinie). Bonn/Deutschland (2018). https://www.bsi.bund.de/DE/Publikationen/TechnischeRichtlinien/tr02102/index_htm.html;jsessionid=D4F0ACAD39ED0893ECBE3F951AE6B66C.2_cid360

[BSI3] Bundesamt für Sicherheit in der Informationstechnik: Sichere Nutzung von WLAN (Technische Richtlinie). Bonn/Deutschland (2018). https://www.bsi.bund.de/SharedDocs/Downloads/DE/BSI/Internetsicherheit/isi_wlan_leitlinie.pdf?__blob=publicationFile

[BSI4] Bundesamt für Sicherheit in der Informationstechnik: Der elektronische Reisepass (ePass) (Internet-Information). https://www.bsi.bund.de/DE/Themen/DigitaleGesellschaft/ElektronischeIdentitaeten/ePass/ePassSeite.html. Accessed 10 Apr 2019

[BSI5] Bundesamt für Sicherheit in der Informationstechnik: Advanced Security Mechanisms for Machine Readable Travel Documents and eIDAS token – Part 2: Protocols (Technische Richtlinie). Bonn/Deutschland (2016). https://www.bsi.bund.de/DE/Publikationen/TechnischeRichtlinien/tr03110/index_htm.html?nn=6615602

O. Manz, *Encrypt, Sign, Attack*, Mathematics Study Resources,
https://doi.org/10.1007/978-3-662-66015-7

[BSI6] Bundesamt für Sicherheit in der Informationstechnik: Advanced Security Mechanisms for Machine Readable Travel Documents and eIDAS token – Part 3: Specifications (Technische Richtlinie). Bonn/Deutschland (2016). https://www.bsi.bund.de/DE/Publikationen/TechnischeRichtlinien/tr03110/index_htm.html?nn=6615602

[BDB] Bundesverband Deutscher Banken: Financial Transaction Services FinTS (Security-Spezifikation). Berlin/Deutschland (2014). https://www.hbci-zka.de/

[DIM] DI-Management: Public key cryptography using discrete logarithms (Internet-Tutorium). https://www.di-mgt.com.au/public-key-crypto-discrete-logs-4-dsa.html. Accessed 10 Apr 2019

[DSB] Datenschutzbeauftragter.: Bitcoin – Technische Grundlagen der Kryptowährung (Internet-Information). https://www.datenschutzbeauftragter-info.de/bitcoin-technische-grundlagen-der-kryptowaehrung/. Accessed 10 Apr 2019

[DeM] Deutsches Museum: Die Rotor-Chiffriermaschine Enigma der deutschen Wehrmacht (Internet-Information). https://www.deutsches-museum.de/sammlungen/meisterwerke/meisterwerke-ii/enigma/. Accessed 10 Apr 2019

[EMV1] EMVCo: Overview (Internet-Information). https://www.emvco.com/about/overview/. Accessed 10 Apr 2019

[EMV2] EMVCo: ICC Specifications for Payment Systems – Security and Key Management (Technische Spezifikation). Foster City CA/USA (2011). https://www.emvco.com/document-search/

[EMV3] EMVCo: ICC Specifications for Payment Systems – Application Specification (Technische Spezifikation). Foster City, CA, USA (2011). https://www.emvco.com/document-search/

[EMV4] EMVCo: Issuer and Application Security Guidelines (Technische Richtlinie). Foster City, CA, USA (2018). https://www.emvco.com/document-search/

[ETC] Enuma Technologies: A Tale of Two Curves (Internet-Information). https://blog.enuma.io/update/2016/11/01/a-tale-of-two-curves-hardware-signing-for-ethereum.html. Accessed 10 Apr 2019

[ETSI1] ETSI: Digital Video Broadcasting (DVB) – Content Scrambling Algorithms (Technische Spezifikation). Sophia Antipolis Cedex, Frankreich (2013). https://www.etsi.org/deliver/etsi_ts/103100_103199/103127/01.01.01_60/ts_103127v010101p.pdf

[ETSI2] ETSI: A5/4 Encription Algorithm for GSM (Technische Spezifikation). Sophia Antipolis Cedex, Frankreich (2011). https://www.etsi.org/deliver/etsi_ts/155200_155299/155226/09.00.00_60/ts_155226v090000p.pdf

[ETSI3] ETSI: MILENAGE Algorithm for UMTS (Technische Spezifikation). Sophia Antipolis Cedex, Frankreich (2010). https://www.etsi.org/deliver/etsi_ts/135200_135299/135206/09.00.00_60/ts_135206v090000p.pdf

[Fox] Fox, D.: Sicherheit des Bluetooth-Standards (Übersichtsartikel). Tagungsband des Deutschen IT-Sicherheitskongresses des BSI, Ingelheim/Deutschland (2003). https://www.secorvo.de/publikationen/bluetooth-sicherheit-fox-2003.pdf

[Fra] Franz, E.: Kryptographie und Kryptoanalyse (Vorlesungsfolien). Dresden, Deutschland (2015). https://www.inf.tu-dresden.de/content/institutes/sya/dud/lectures/2015sommersemester/Kryptoanalyse/KuKA15_01_1s.pdf

[Gar] Gartner, L.: Häufigkeitstabellen (Internet-Blog). http://www.mathe.tu-freiberg.de/~hebisch/cafe/kryptographie/haeufigkeitstabellen.html. Accessed 10 Apr 2019

[HEZ] Hassan, Z., Elgard, T., Zekry, A.: Modifying authentication techniques in mobile communication systems (Forschungsartikel). Int. J. Eng. Res. Appl. **4**, Kairo, Ägypten (2014)

[Hau1] Hauck, P.: Kryptologie und Datensicherheit (Vorlesungsskript). Tübingen, Deutschland (2009)

[Hau2] Hauck, P.: Kryptologie (Vorlesungsskript). Tübingen, Deutschland (2015). https://www.fsi.uni-tuebingen.de/_media/studium/skripte/kryptows1415.pdf

[Hau3] Hauck, P.: Primzahltests und Faktorisierungsalgorithmen (Vorlesungsskript), Tübingen, Deutschland (2007)

[IWS] IWS: Descriptions of SHA-256, SHA-384, and SHA-512 (Technische Spezifikation). (2000). http://www.iwar.org.uk/comsec/resources/cipher/sha256-384-512.pdf

[JuM] Jurisic, A., Menezes, A.: Elliptic Curves and Cryptography (Übersichtsartikel). Alabama, USA (1999). http://www.cs.nthu.edu.tw/~cchen/CS4351/jurisic.pdf

[Kak] Karpfinger, C., Kiechle, H.: Kryptologie (Lehrbuch). Vieweg+Teubner, Wiesbaden (2010)

[Kin] Kingston-Technology: Verschlüsselte USB-Sticks – XTS-Verschlüsselung (Produkt-Information). https://www.kingston.com/de/usb/encrypted_security/xts_encryption. Accessed 10 Apr 2019

[Kob] Koblitz, N.: A Course in Number Theory and Cryptography (Lehrbuch). Springer, New York (1994)

[Kuh] Kuhlemann, O.: Illuminati Code (Internet-Information). http://kryptografie.de/kryptografie/chiffre/illuminati.htm. Accessed 10 Apr 2019

[Lan] Lang, H.W.: Kryptografie für Dummies (Sachbuch). Wiley-VCH, Weinheim (2018)

[LoM] Lochter, M., Merkle, J.: ECC Brainpool Standard: Curces and Curve Generation (Internet-Information). https://tools.ietf.org/html/rfc5639. Accessed 10 Apr 2019

[Man] Manz, O.: Fehlerkorrigierende Codes (Lehrbuch). Springer Vieweg, Wiesbaden (2017)

[Rau] Rau, T.: Verschlüsselung von Passwörtern (Internet-Blog). https://www.herr-rau.de/wordpress/2013/04/kryptographie-1-verschluesselung-von-passwoertern.htm. Accessed 10 Apr 2019

[Rei] ReinerSCT.: Wikibanking: Online-Banking (Internet-Information). http://www.wikibanking.net/onlinebanking/verfahren/hbci/. Accessed 10 Apr 2019

[Scw] Schwenk, J.: Systemsicherheit – Pay-TV (Vorlesungsfolien). Bochum/Deutschland (2004). http://www.ruhr-uni-bochum.de/nds/lehre/vorlesungen/systemsicherheit_alt/Systemsicherheit_4_2B.pdf

[SEC] Certicom Corp: Standards for Efficient Cryptography, Recommended Elliptic Curve Domain Parameters (Specification). (2010). http://www.secg.org/sec2-v2.pdf

[Sto] Stockinger, T.: GSM Network and Its Privacy – The A5 Stream Cipher (Übersichtsartikel). (2005). http://citeseerx.ist.psu.edu/viewdoc/download?doi=10.1.1.465.8718&rep=rep1&type=pdf

[TeM] Teunissen, P., Montenbruck, O.: Springer Handbook of Global Navigation Satellite Systems (Fachbuch). Springer, Cham (2017)

[USG] U.S. Government (USG): Mathematical routines for the NIST prime elliptic curves (Technische Spezifikation). (2010). http://citeseerx.ist.psu.edu/viewdoc/download?doi=10.1.1.204.9073&rep=rep1&type=pdf

[WhA] WhatsApp: Encryption Overview (Technical White Paper). (2017). https://www.whatsapp.com/security/WhatsApp-Security-Whitepaper.pdf

[WPAES] Wikipedia: Advanced Encryption Standard (Internet-Enzyklopädie). https://en.wikipedia.org/wiki/Advanced_Encryption_Standard. Accessed 10 Apr 2019

[WPA5A] Wikipedia: A5 (Algorithmus) (Internet-Enzyklopädie). https://de.wikipedia.org/wiki/A5_(Algorithmus). Accessed 10 Apr 2019

[WPBGS] Wikipedia: Baby-Step-Giant-Step-Algorithmus (Internet-Enzyklopädie). https://de.wikipedia.org/wiki/Babystep-Giantstep-Algorithmus. Accessed 10 Apr 2019

[WPBeM] Wikipedia: Betriebsmodus (Internet-Enzyklopädie). https://de.wikipedia.org/wiki/Betriebsmodus_(Kryptographie). Accessed 10 Apr 2019

[WPBiC] Wikipedia: Bitcoin (Internet-Enzyklopädie). https://de.wikipedia.org/wiki/Bitcoin. Accessed 10 Apr 2019

[WPBlC] Wikipedia: Blockchain (Internet-Enzyklopädie). https://de.wikipedia.org/wiki/Blockchain. Accessed 10 Apr 2019

[WPBlu] Wikipedia: Bluetooth (Internet-Enzyklopädie). https://de.wikipedia.org/wiki/Bluetooth. Accessed 10 Apr 2019

[WPCSA] Wikipedia: Common Scrambling Algorithm (Internet-Enzyklopädie). https://en.wikipedia.org/wiki/Common_Scrambling_Algorithm. Accessed 10 Apr 2019

[WPCrC] Wikipedia: Cryptocurrency (Internet-Enzyklopädie). https://en.wikipedia.org/wiki/Cryptocurrency. Accessed 10 Apr 2019

[WPC29] Wikipedia: Curve25519 (Internet-Enzyklopädie). https://de.wikipedia.org/wiki/Curve25519. Accessed 10 Apr 2019

[WPDES] Wikipedia: Data Encryption Standard (Internet-Enzyklopädie). https://de.wikipedia.org/wiki/Data_Encryption_Standard. Accessed 10 Apr 2019

[WPDKA] Wikipedia: Differenzielle Kryptoanalyse (Internet-Enzyklopädie). https://de.wikipedia.org/wiki/Differenzielle_Kryptoanalyse. Accessed 10 Apr 2019

[WPDHS] Wikipedia: Diffie-Hellman-Schlüsselaustausch (Internet-Enzyklopädie). (geöffnet 10. April 2019). https://de.wikipedia.org/wiki/Diffie-Hellman-Schl%C3%BCsselaustausch

[WPDiC] Wikipedia: Digital currency (Internet-Enzyklopädie). https://en.wikipedia.org/wiki/Digital_currency. Accessed 10 Apr 2019

[WPDSA] Wikipedia: Digital Signature Algorithm (Internet-Enzyklopädie). https://en.wikipedia.org/wiki/Digital_Signature_Algorithm. Accessed 10 Apr 2019

[WPDET] Wikipedia: Disk encryption theory (Internet-Enzyklopädie). https://en.wikipedia.org/wiki/Disk_encryption_theory. Accessed 10 Apr 2019

[WPECC] Wikipedia: Elliptic-curve cryptography (Internet-Enzyklopädie). https://en.wikipedia.org/wiki/Elliptic-curve_cryptography. Accessed 10 Apr 2019

[WPECD] [Wikipedia: Elliptic Curve DSA (Internet-Enzyklopädie). https://de.wikipedia.org/wiki/Elliptic_Curve_DSA. Accessed 10 Apr 2019

[WPEKu] Wikipedia: Elliptische Kurve (Internet-Enzyklopädie). https://de.wikipedia.org/wiki/Elliptische_Kurve. Accessed 10 Apr 2019

[WPEnM] Wikipedia: Enigma (Maschine) (Internet-Enzyklopädie). https://de.wikipedia.org/wiki/Enigma_(Maschine). Accessed 10 Apr 2019

[WPFPV] Wikipedia: Festplattenverschlüsselung (Internet-Enzyklopädie). https://de.wikipedia.org/wiki/Festplattenverschl%C3%BCsselung. Accessed 10 Apr 2019

[WPGPS] Wikipedia: Global Positioning System (Internet-Enzyklopädie). https://de.wikipedia.org/wiki/Global_Positioning_System. Accessed 10 Apr 2019

[WPGPG] Wikipedia: GNU Privacy Guard (Internet-Enzyklopädie). https://en.wikipedia.org/wiki/GNU_Privacy_Guard. Accessed 10 Apr 2019

[WPHMA] Wikipedia: Keyed-Hash Message Authentication Code (Internet-Enzyklopädie). https://de.wikipedia.org/wiki/Keyed-Hash_Message_Authentication_Code. Accessed 10 Apr 2019

[WPHBC] Wikipedia: Homebanking Computer Interface (Internet-Enzyklopädie). https://de.wikipedia.org/wiki/Homebanking_Computer_Interface. Accessed 10 Apr 2019

[WPIPF] Wikipedia: Internetprotokollfamilie (Internet-Enzyklopädie). https://de.wikipedia.org/wiki/Internetprotokollfamilie. Accessed 10 Apr 2019

[WPKAS] Wikipedia: KASUMI (Internet-Enzyklopädie). https://de.wikipedia.org/wiki/KASUMI. Accessed 10 Apr 2019

[WPKrA] Wikipedia: Kryptoanalyse (Internet-Enzyklopädie). https://de.wikipedia.org/wiki/Kryptoanalyse. Accessed 10 Apr 2019

[WPLvK] Wikipedia: Liste von Kryptowährungen (Internet-Enzyklopädie). https://de.wikipedia.org/wiki/Liste_von_Kryptow%C3%A4hrungen. Accessed 10 Apr 2019

[WPMD6] Wikipedia: MD6 (Internet-Enzyklopädie). https://en.wikipedia.org/wiki/MD6. Accessed 10 Apr 2019

[WPMMV] Wikipedia: Merkles Meta-Verfahren (Internet-Enzyklopädie). https://de.wikipedia.org/wiki/Merkles_Meta-Verfahren. Accessed 10 Apr 2019

[WPMRP] Wikipedia: Miller-Rabin primality test (Internet-Enzyklopädie). https://en.wikipedia.org/wiki/Miller%E2%80%93Rabin_primality_test. Accessed 10 Apr 2019

[WPOCF] Wikipedia: One-way compression function (Internet-Enzyklopädie). https://en.wikipedia.org/wiki/One-way_compression_function. Accessed 10 Apr 2019

[WPOPG] Wikipedia: OpenPGP (Internet-Enzyklopädie). https://de.wikipedia.org/wiki/OpenPGP. Accessed 10 Apr 2019

[WPPRL] Wikipedia: Pollard's rho algorithm for logarithms (Internet-Enzyklopädie). https://en.wikipedia.org/wiki/Pollard%27s_rho_algorithm_for_logarithms. Accessed 10 Apr 2019

[WPPSK] Wikipedia: Pre-shared Key (Internet-Enzyklopädie). https://de.wikipedia.org/wiki/Pre-shared_key. Accessed 10 Apr 2019

[WPPGP] Wikipedia: Pretty Good Privacy (Internet-Enzyklopädie). https://de.wikipedia.org/wiki/Pretty_Good_Privacy. Accessed 10 Apr 2019

[WPQSi] Wikipedia: Quadratisches Sieb (Internet-Enzyklopädie). https://de.wikipedia.org/wiki/Quadratisches_Sieb. Accessed 10 Apr 2019

[WPRAD] Wikipedia: Remote Authentication Dial-In User Service (Internet-Enzyklopädie). https://de.wikipedia.org/wiki/Remote_Authentication_Dial-In_User_Service. Accessed 10 Apr 2019

[WPSH2] Wikipedia: SHA-2 (Internet-Enzyklopädie). (geöffnet 10. April 2019). https://de.wikipedia.org/wiki/SHA-2

[WPSH2e] Wikipedia: SHA-2 engl. (Internet-Enzyklopädie). https://en.wikipedia.org/wiki/SHA-2. Accessed 10 Apr 2019

[WPSH3] Wikipedia: SHA-3 (Internet-Enzyklopädie). https://de.wikipedia.org/wiki/SHA-3. Accessed 10 Apr 2019

[WPStV] Wikipedia: Stromverschlüsselung (Internet-Enzyklopädie). https://de.wikipedia.org/wiki/Stromverschl%C3%BCsselung. Accessed 10 Apr 2019

[WPTLS] Wikipedia: Transport Layer Security (Internet-Enzyklopädie). https://de.wikipedia.org/wiki/Transport_Layer_Security. Accessed 10 Apr 2019

[WPTLSe] Wikipedia: Transport Layer Security engl. (Internet-Enzyklopädie). https://en.wikipedia.org/wiki/Transport_Layer_Security. Accessed 10 Apr 2019

[WPWLA] Wikipedia: Wireless Local Area Network (Internet-Enzyklopädie). https://de.wikipedia.org/wiki/Wireless_Local_Area_Network. Accessed 10 Apr 2019

[WPWP2] Wikipedia: WPA2 (Internet-Enzyklopädie). https://de.wikipedia.org/wiki/WPA2. Accessed 10 Apr 2019

[WPZBS] Wikipedia: Zugangsberechtigungssystem (Internet-Enzyklopädie). https://de.wikipedia.org/wiki/Zugangsberechtigungssystem. Accessed 10 Apr 2019

[WPZIP] Wikipedia: ZIP-Dateiformat (Internet-Enzyklopädie). https://de.wikipedia.org/wiki/ZIP-Dateiformat. Accessed 10 Apr 2019

[WZi] Winzip: AES: A More Secure Method of Encryption (Produktinformation). https://www.winzip.com/learn/aes-encryption.html. Accessed 10 Apr 2019

[Wil] Willems, W.: Codierungstheorie und Kryptographie (Lehrbuch). Birkhäuser-Springer, Basel (2008)

Index

O. Manz, *Encrypt, Sign, Attack*, Mathematics Study Resources,
https://doi.org/10.1007/978-3-662-66015-7